T0288893

Effects of Bonuses on Active Component Reenlistment Versus Prior Service Enlistment in the Selected Reserve

James Hosek, Trey Miller

Prepared for the Office of the Secretary of Defense

NATIONAL DEFENSE RESEARCH INSTITUTE

The research described in this report was prepared for the Office of the Secretary of Defense (OSD). The research was conducted within the RAND National Defense Research Institute, a federally funded research and development center sponsored by OSD, the Joint Staff, the Unified Combatant Commands, the Navy, the Marine Corps, the defense agencies, and the defense Intelligence Community under Contract W74V8H-06-C-0002.

Library of Congress Control Number: 2011929826

ISBN: 978-0-8330-5216-2

The RAND Corporation is a nonprofit institution that helps improve policy and decisionmaking through research and analysis. RAND's publications do not necessarily reflect the opinions of its research clients and sponsors.

RAND® is a registered trademark.

Cover photo: U.S. Army photo by Spc. Breanne Pye

Published 2011 by the RAND Corporation
1776 Main Street, P.O. Box 2138, Santa Monica, CA 90407-2138
1200 South Hayes Street, Arlington, VA 22202-5050
4570 Fifth Avenue, Suite 600, Pittsburgh, PA 15213-2665
RAND URL: http://www.rand.org/
To order RAND documents or to obtain additional information, contact
Distribution Services: Telephone: (310) 451-7002;
Fax: (310) 451-6915; Email: order@rand.org

Preface

During the past two decades, the reserves have been called upon repeatedly to take part in the nation's military operations. In earlier years, the role of the reserves had been that of a strategic force to be engaged only under extraordinary circumstances, such as major theater war. The transformation of the reserves into a force both strategic and operational has demonstrated the capability of reservists to participate in extended military operations and brought new importance to reserve readiness. The concept of a continuum of service, efforts to forge congruence between reserve and active personnel policies and compensation structure, the commitment to provide reserve units with up-to-date equipment, and the effort to support families of deployed reservists are all aspects of the transformation. These changes have sparked interest in the reserves, and policymakers are becoming more conversant with reserve force structure, missions, and policy and are seeking more information about the effectiveness of reserve policy instruments.

A major source of reserve manpower is the flow of enlisted members from an active component (AC) to a reserve component (RC). The initial focus of the research in this volume was determining how effective RC bonuses are in attracting these prior service members. However, both AC and RC bonuses affect AC service members' decision to reenlist in the AC, join the RC, or become a civilian. The focus of the research thus broadened from just the effectiveness of RC bonuses to how both AC and RC bonuses interact to affect both AC reenlistment and prior service enlistment in the RC. The research findings may be of interest to policymakers and defense officials with responsibility for ensuring that the AC and RC are fully manned, such as congressional committees on military personnel, military commanders, personnel force planners and programmers, recruiting commanders, retention counselors, bonus allocators, and military compensation officials. The research may also interest defense manpower analysts.

This research was sponsored by the Office of the Secretary of Defense for Reserve Affairs and conducted within the Forces and Resources Policy Center of the RAND National Defense Research Institute, a federally funded research and development center sponsored by the Office of the Secretary of Defense, the Joint Staff, the Unified Combatant Commands, the Navy, the Marine Corps, the defense agencies, and the defense Intelligence Community.

For more information on the RAND Forces and Resources Policy Center, see http://www.rand.org/nsrd/ndri/centers/frp.html or contact the director (contact information is provided on the web page).

Contents

Figures

Tables

Summary

The reserve components should be fully manned and ready, but at times during 2004 to 2009 the Army National Guard, the Army Reserve, and the Marine Corps Reserve experienced manning shortfalls. The shortfalls occurred despite constant manpower authorizations and can be attributed to an insufficient supply of personnel. Supply in general depends on recruiting and retention, and retention was stable in the Army National Guard and the Marine Corps Reserve, though it did decrease in the Army Reserve. Thus, in the Army National Guard, the Marine Corps Reserve, and to some extent the Army Reserve the manning shortfall came from an inadequate inflow of recruits.

The present study began with a focus on the effectiveness of RC enlistment and affiliation bonuses in increasing the enlistment of service members leaving the AC. In our analysis, we assume that AC service members at a reenlistment point consider remaining in the AC, joining the RC, or leaving the military entirely, and that both AC and RC bonuses influence this decision. Framing the analysis in this way broadened the focus of the study from just the effectiveness of RC bonuses to include how AC and RC bonuses interact to affect both AC reenlistment and prior service enlistment in the RC.

We focus on several research questions:

- Are RC bonuses effective in attracting AC enlisted members?
- How do the effects of AC and RC bonuses interact? That is, to what extent do RC enlistment bonuses draw service members away from the AC, and to what extent to do AC reenlistment bonuses reduce prior service enlistment in the RC?
- What are the total and incremental costs of RC bonuses as bonus levels change?

In approaching these questions, we first discuss trends in reserve strength and RC and AC bonus programs to provide policy context for the study. We describe our data and present a theoretical model of the AC/RC/civilian choice, and we then discuss the empirical estimates of bonus effects, deployment effects, and time effects. Next, we develop models of bonus-setting behavior and present estimates of bonus total and incremental cost based on the empirical estimates. In the final chapter, we present our conclusions.

Trends in Reserve and Active Component Bonus Programs

The year 2004 marked a low point in the use of AC and RC bonuses. Saddam Hussein's army was defeated in spring 2003, and retention in the U.S. military was high. But in subsequent months, insurgent groups became active and conditions worsened. The United States responded with a high ongoing pace of deployments, and many service members had two or even three deployments during this period. By 2005, the Army and the Marine Corps experienced downward pressure from deployment on retention (Hosek and Martorell, 2009), and the Army National Guard and Army Reserve coped with shortfalls in their personnel strength relative to authorized levels. The AC Army and Marine Corps greatly expanded the number of specialties eligible for a reenlistment bonus and increased average bonus amounts in 2005. These services also set plans in motion to grow, as Congress mandated. Bonus usage decreased in the AC Navy and Air Force as these services downsized, though bonus amounts increased among those receiving bonuses.

RC bonus usage and amounts were quite small up to 2006. In 1999, 11 percent of Army personnel who transitioned from the AC to the RC received a bonus, and that number was lower for the other services. Moreover, bonuses were small and averaged less than $2,000 for all services. RC bonus eligibility criteria were broadened and bonus ceilings were raised in 2006, paving the way to an expansion of the RC bonus program. Much of the expansion occurred through increasing average bonus amounts. The average RC bonus increased from about $1,500 to more than $10,000 from 2006 to 2009, though the number of RC bonuses awarded changed little.

The large changes in AC and RC bonus usage and amounts during these years provide variation helpful in identifying the impact of bonuses on AC reenlistment and RC prior service enlistment.

Data and Model

Our analysis comprises the universe of all reenlistment decision points reached by active-duty military personnel with three or more years of service between 1999 and 2008. The analysis file is built from Defense Manpower Data Center administrative files on personnel. Data on bonus offers are not available, so we impute AC and RC bonuses to each service member. The imputations are made by cells defined by three-digit AC occupational specialty, zone (year of service group), month, and AC component. The imputed AC bonus is based on bonus payments received by those service members in the cell who chose to reenlist. The imputed RC bonus is based on those service members in the cell who chose to join the RC within two years after leaving AC and who received a bonus.[1]

Our model assumes that a service member at a reenlistment decision point chooses among three alternatives—AC reenlistment, RC enlistment, and becoming a civilian—and selects the alternative with the highest utility. The utility of AC reenlistment depends on the AC bonus, the utility of RC enlistment depends on the RC bonus, and the utility of each alternative also depends on other variables, such as demographics, and an error term. We assume the error follows a generalized extreme value distribution, which leads to the conditional logit model of choice probabilities. The model provides a framework for estimating the effects of the AC and RC bonuses.

The estimation of bonus effects presents issues of possible biases stemming from reverse causation,[2] omitted variables, bonus caps, the assumption of a static decisionmaker,[3] and the unavailability of vacancies in particular units. We control for these issues to some extent by including variables for occupational specialty and year and by employing a quadratic specification of the bonus. Variables in the model also include Armed Forces Qualification Test (AFQT) category, education, gender, race, and an indicator for being promoted faster than average.

Bonus Effects

We estimate the model for service members in each AC component at first-term and second-term-or-higher reenlistment decision points. We find positive, statistically significant AC and RC bonus impacts. We use the estimated bonus coefficients to project AC reenlistment rates and RC enlistment rates for various combinations of AC and RC bonuses. The projections indicate that increasing the AC bonus increases AC reenlistment and decreases RC enlistment, while increasing the RC bonus increases RC enlistment and decreases AC reenlistment. That is, the own-effect of bonuses is positive and the cross-effect of bonuses is negative. The magnitude of these effects varies by term and service. The bonus specification is quadratic, and incremental bonus impacts decrease as bonus amounts reach higher levels. The change in incremental impact by bonus amount is small for the AC but more noticeable for the RC.

The general pattern of positive own-effects and smaller, negative cross-effects of bonuses on first-term reenlistment decisions is the same across services, but the mag-

---

[1]   Nearly all AC service members who join the RC do so within two years of leaving the AC.

[2]   Low RC enlistment can cause the RC bonus to be increased, whereas we are interested in the effect of a higher RC bonus on RC enlistment.

[3]   Congress sets a maximum bonus under each of the bonus programs. We discuss reasons why the existence of these bonus caps may bias our estimates.

nitudes differ. We find somewhat larger own-effects of RC bonuses for the Navy and Marine Corps. The cross-effect of AC bonuses on RC enlistment rates is roughly constant across services, but the cross-effect of RC bonuses on AC reenlistment rates is somewhat stronger for the Air Force and somewhat weaker for the Marine Corps.

The patterns for AC members at the second term and higher are similar to those for first-term members. An exception is that we find no evidence that RC bonuses affect second-term-and-higher reenlistment decisions for the Marine Corps. We also checked whether bonus effects vary by whether the service member's occupational specialty was classified as combat arms and by the amount and type of deployment while on active duty, but we found little evidence of interaction effects.

Bonus-Setting Behavior

The empirical finding of negative bonus cross-effects has implications for bonus setting and bonus costs that we delineate through models of AC and RC bonus setting. In the models, each side has an objective, e.g., meeting a reenlistment or enlistment target, and reacts to the bonus action taken by the other as needed to meet the target.

We present models for two bonus-setting objectives: (1) reaching specified targets for AC reenlistments and RC enlistments and (2) maximizing the value of additional AC reenlistments and RC enlistments relative to their bonus cost. The targets in the first case are given, and the model raises the question of how the targets are determined. In the second case, we assume that bonus setting is determined by an assessment of the perceived value of an additional AC reenlistment or RC enlistment versus the additional cost.

In the first case, bonus setters set bonuses at the minimum value necessary to achieve their target, conditional on the other component's bonus. This produces interactive behavior as bonuses are changed and leads to an equilibrium in which AC and RC bonuses depend only on the AC and RC targets. In the second case, optimal bonuses again depend on the interactive behavior and are such that the incremental value of an additional AC reenlistment equals the incremental cost, and similarly for an additional RC enlistment.

Under either objective, a bonus increase creates a negative externality for the other component. This is a consequence of supply-side behavior and is not a result of some inefficiency in bonus setting; a service member's willingness to reenlist in the AC depends positively on the AC bonus and negatively on the RC bonus, and the willingness to join the RC depends positively on the RC bonus and negatively on the AC bonus. Because of the negative externality, the other component must increase its bonus, which causes the cost of its bonus program to increase. Coordination between AC and RC bonus setters may help to avoid surprises but will not eliminate the cross-effects, because they derive from supply behavior. To reiterate, the need to adjust the

RC bonus when the AC bonus is changed, and vice versa, is a rational response required by the supply-side behavior and does not represent waste or inefficiency. Nevertheless, coordination between the AC and RC can help to make the process of setting bonuses and budgeting for bonus programs as smooth as possible.

Bonus Cost

Combining insights from the bonus-setting models with the empirical estimates of bonus impacts, we derive estimates for the cost of AC and RC bonuses from two different perspectives. The first shows AC and RC bonus costs for different combinations of AC and RC bonus amounts. While intuitive, this calculation ignores the externality that occurs when a bonus increase in one component decreases the supply of service members to the other component. The second perspective overcomes this limitation by showing the AC and RC bonus amounts required to sustain different combinations of AC reenlistment and RC enlistment rates, and the corresponding bonus costs. Using this approach, the bonus cost has two components: (1) the cost of a component's higher bonus and (2) the cost of the other component's higher bonus needed to hold its rate constant.

At an AC bonus of $8,000 and an RC bonus of $2,000, the first perspective indicates an average cost per RC enlistment ranging from $28,000 for the Marine Corps to $40,000 for the Army, and an average cost per AC reenlistment of $22,000 for the Marine Corps to $54,000 for the Army and $132,000 for the Air Force. The Air Force estimate is high because of low responsiveness to reenlistment bonuses among Air Force personnel, a finding consistent with past research (Hosek and Martorell 2009; Asch et al., 2010).

As an example of the second perspective, a one-point increase in the Navy RC enlistment rate from 0.155 to 0.165 at an AC reenlistment rate of 0.40 requires an increase in the RC bonus from $1,429 to $3,755 and an increase in the AC bonus from $10,656 to $11,265. The RC bonus cost per 100 AC members at reenlistment increases from $22,000 to $62,000, or by $40,000, and the AC bonus cost increases from $426,000 to $451,000, or by $25,000. Again, these cross-effect costs are rational responses required because of supply-side behavior. They do not represent wastage and are part of the full cost of a bonus increase.

Acknowledgments

We would like to thank LTC Richard Dederer and Tom Liuzzo of the Office of the Secretary of Defense, Reserve Affairs, as well as LTC Richard Cardenas and John Winkler, both formerly of Reserve Affairs, for their support. We received valuable comments from many RAND colleagues, including Beth Asch, Michael Hansen, Heather Krull, Tom Lippiatt, Paco Martorell, Michael Mattock, and Sebastian Negrusa, and from Colin Doyle of the Institute for Defense Analyses and other participants at the 2010 Western Economics Association conference. We appreciate the database and programming support provided by Suzy Adler, Arthur Bullock, Craig Martin, and Laurie McDonald. The report benefited considerably from reviews by Paul Heaton of RAND and Paul Hogan of Lewin Associates, and we thank them.

Abbreviations

AC	active component
AFQT	Armed Forces Qualification Test
ANG	Air National Guard
ARNG	Army National Guard
DMDC	Defense Manpower Data Center
DoD	U.S. Department of Defense
ETS	expiration of term of service
GEV	generalized extreme value
IIA	independence of irrelevant alternatives
JAMRS	Joint Advertising and Marketing Research System
JUMPS	Joint Uniform Military Pay System
MOS	military occupation specialty
MSO	military service obligation
OLS	ordinary least squares
RC	reserve component
RPF	Reserve Pay File
USAFR	U.S. Air Force Reserve
USAR	U.S. Army Reserve
USMCR	U.S. Marine Corps Reserve
USNR	U.S. Navy Reserve

CHAPTER ONE

Introduction

This study began with the goal of learning about factors affecting the flow of enlisted personnel from an active component (AC) of the U.S. military to a reserve component (RC), and in particular of learning about the effectiveness of reserve enlistment and affiliation bonuses. However, since both AC and RC bonuses affect AC service members' decision to reenlist in the AC, join the RC, or become a civilian, we examine how both AC and RC bonuses interact to affect both AC reenlistment and prior service enlistment in the RC. We focus on the following questions:

- Are RC bonuses effective in attracting AC enlisted members?
- How do the effects of AC and RC bonuses interact? That is, to what extent do RC enlistment bonuses draw service members away from the AC, and to what extent to do AC reenlistment bonuses reduce prior service enlistment in the RC?
- What are the total and incremental costs of RC bonuses as bonus levels change?

These questions have become increasingly relevant because of the growing reliance on RC forces in national security. Today's RC is strategic and operational. Because the RC is called upon to train for and deploy to peacekeeping, humanitarian, anti-terror, and anti-insurgent military operations and to take part in homeland security and border operations, having RC forces that are fully manned is critically important. Yet during 2004 to 2009, the RC had manning shortfalls in the Army National Guard, Army Reserve, and Marine Corps Reserve.[1] For the most part, the shortfalls were not the result of an increase in authorized billets (i.e., the quantity of personnel demanded) or a decrease in retention, but rather came from too few recruits.

Our study concerns a major source of RC recruits—prior service recruits, i.e., individuals who initially serve in the AC—and we are interested in the effectiveness of two flexible recruiting tools, RC enlistment and affiliation bonuses. Bonuses have proved useful in the AC for many years, and studies based on administrative data and field experiments have helped to quantify their effectiveness and cost-effectiveness. In contrast, the research questions posed above have not been answered.

---

[1] These shortfalls are discussed in Chapter Two.

The past literature on reserve recruiting is scant. Kostiak and Grogan (1987) find that monthly state-level enlistments into the Navy Reserve are positively related to the number of recruiters and to the state unemployment rate and population, and negatively related to civilian pay relative to military pay. Their study does not distinguish between prior and nonprior service recruits or include bonuses. Hattiangadi et al. (2006) tabulate the percentage of marines who join the Selected Reserve after leaving the AC Marine Corps, and Asch and Loughran (2005) provide a qualitative assessment of whether a more generous education benefit would help in meeting high-quality prior and nonprior service RC recruiting goals. These studies do not model prior service enlistment into the RC or seek to estimate bonus effects. Arkes and Kilburn (2005) estimate models of prior and nonprior service RC enlistment, though their variables do not include bonuses and they describe their prior service results as "unreliable." Their prior service analysis focuses on individuals who left the AC and are civilians. In contrast, our analysis focuses on AC members at a reenlistment decision point who face the choice of reenlistment, enlistment into the RC, or leaving the military entirely, and our analysis includes AC and RC bonuses.

To establish policy context, Chapter Two discusses RC manpower authorizations (funded spaces) and inventory, which together indicate where shortages exist. Chapter Two also describes AC and RC bonus programs, AC reenlistment rates, and rates of transition to the RC. Chapter Three discusses our data and model of choice behavior and identifies issues in the estimation of bonus effects that are discussed more fully in Appendix B. Chapter Four presents AC and RC bonus effect estimates as well as results for deployment and year fixed-effect variables. Chapter Five considers bonus setting and presents estimates of bonus costs from two perspectives that illustrate the interplay of AC and RC bonuses with respect to bonus amount and cost. Chapter Six offers closing thoughts.

CHAPTER TWO

Background

In this chapter, we describe RC enlistment and affiliation bonuses and review the expansion in the generosity of these bonuses implemented in 2006. Because we want to determine whether these bonuses were effective in attracting AC members at a reenlistment decision point, we also discuss AC reenlistment bonuses. The discussion provides information about transition rates from the AC to the RC, active reenlistment rates and bonus usage, reserve enlistment rates among AC members at a reenlistment point, and reserve bonus usage.

The aggregate data in this chapter indicate that the Army and Marine Corps faced the greatest pressure in meeting their RC recruiting requirements and therefore are of particular interest in our empirical analysis. Moreover, the manning shortfalls relative to authorized positions experienced by the reserve components of the Army and the Marine Corps chiefly result from a decrease in supply rather than an increase in authorizations or a decrease in retention.

Several factors may have affected supply. The civilian labor market had a low and declining unemployment rate, making it relatively easy to find a civilian job. By the middle of the decade, the military operations in Iraq and Afghanistan showed no sign of abating, and the use of improvised explosive devices (IEDs) was widespread. Many AC members had been deployed at least once and faced more deployment, whether they stayed in the AC or transitioned to the RC. For some, the prospect of more deployment could have deterred them from staying in the military. AC bonus usage increased, helping to increase AC reenlistment but perhaps decreasing the flow of personnel to the RC. In Chapter Four, we present estimates of the effect of AC bonuses and AC deployment on RC enlistment. We also find evidence of a downward trend in joining the RC from the AC after controlling for other factors, a trend that might reflect a reaction to expectations of future deployment and their dangers. The decrease in RC Army and Marine Corps supply is notable because RC bonuses are a policy tool for increasing supply, and we are especially interested in whether RC bonuses proved effective at increasing RC prior service enlistment rates.

Trends in the Selected Reserves

The reserve forces include the Individual Ready Reserve (IRR), the Retired Reserve, the Selected Reserve, and, as part of the latter, the Active Guard and Reserve (AGR). The Selected Reserve is composed of the Army National Guard (ARNG), the U.S. Army Reserve (USAR), the U.S. Navy Reserve (USNR), the U.S. Marine Corps Reserve (USMCR), the Air National Guard (ANG), and the U.S. Air Force Reserve (USAFR). Each of these components is made up of units, whereas the Individual Ready Reserve and the Retired Reserve are composed of individuals who are not attached to Selected Reserve units. When the reserves are used in a military operation, it is a Selected Reserve unit that is deployed, and, in effect, the Selected Reserves are structured and supported to be capable and ready for use in military operations, given appropriate time to prepare.

Table 2.1 shows the number of Selected Reserve personnel and authorized strength for officers and enlisted in April 2010. Reserve strength totaled over 850,000, and every component was at or above authorized strength. The Army National Guard and the U.S. Army Reserve together had a strength of 570,000, the Air National Guard and U.S. Air Force Reserve had a combined strength of 177,000, the U.S. Naval Reserve strength was 65,000, and the U.S. Marine Corps Reserve strength was just under 40,000. Overall, 17 percent of the Selected Reserves were officers, about the same as the percentage in the AC. The Marine Corps Reserves and the Army National Guard had the lowest percentage of officers (11 percent and 13 percent, respectively), while the Navy Reserve and Air Force Reserve had the highest percentage (28 percent and 27 percent).

Figure 2.1 shows the enlisted length of service distribution in April 2010, including years of AC service, if any. At that time, 40 percent had a length of service of four years or fewer, 50 percent had six or fewer years, and two-thirds had 10 or fewer years.

Table 2.1
Selected Reserve Strength and Authorizations, April 2010

Reserve Component	Officer	Enlisted	Total	Percentage of Officers	Authorized	Total as Percentage of Authorized
ARNG	40,964	321,978	362,942	13	358,200	101
USAR	36,475	171,497	207,972	21	205,000	101
USNR	14,224	51,269	65,493	28	65,500	100
USMCR	3,769	35,690	39,459	11	39,600	100
ANG	14,416	94,102	108,518	15	106,700	102
USAFR	14,653	54,620	69,273	27	69,500	100
DoD total	124,501	729,156	853,657	17	844,500	101

SOURCE: Defense Manpower Data Center.

Figure 2.1
Selected Reserve Enlisted Length of Service, April 2010

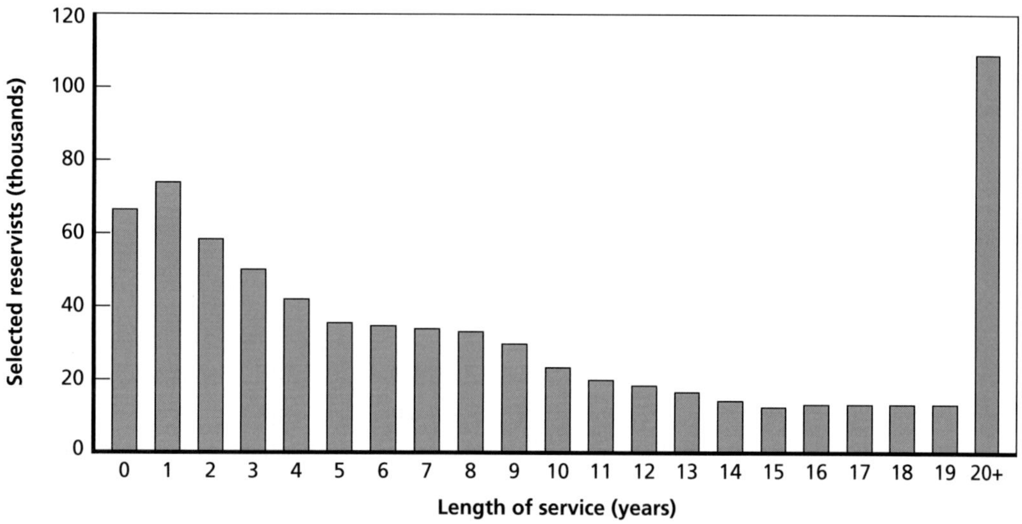

NOTE: The figure puts reservists with 20 or more years into a single category, "20+," but this should not be taken to mean that there is a spike of personnel at the 20-year point.
RAND *MG1057-2.1*

Fifteen percent had 20 or more years. The distribution for the reserves is roughly similar to that of the actives, though with more past 20 years.

It is likely that many reservists in year-of-service zero joined the reserves without prior service in the AC. In April 2010, the year-zero group made up 9 percent of the Selected Reservists, and outflow from the reserves ("attrition") averaged 19 percent from 2005 to 2009. These figures suggest that nonprior service enlistees are about half of all enlistees, with the other half coming from members who first served in an active component, plus those who joined the reserves directly, then left, and then reenlisted. Both those coming from the AC and those reentering the reserves, but not having been in an active component, would enter the reserves at a length of service greater than zero.

Although personnel strength was at its authorized level in April 2010, it has not always been so in the past decade. The Army Guard and Reserve had shortfalls during 2004–2007, and the Marine Corps Reserve had shortfalls in 2007–2009. Table 2.2, covering 1998–2010, has four panels. The first and second panels present strength and authorization by component, the third panel shows strength as a percentage of authorization, and the fourth panel shows the numerical surplus or shortfall.

The Army National Guard authorization was very close to constant at 350,000 from 2000 to 2009. Strength exceeded authorization in these years, except in 2004–2006. The worst shortfall, in 2005, was nearly 17,000. A shortfall results from too much outflow or too little inflow relative to authorized strength, and below we present

Table 2.2
Selected Reserve Strength and Authorization, by Component, 1998–2010

	1998	1999	2000	2001	2002	2003	2004	2005	2006	2007	2008	2009	2010
Strength													
ARNG	362,444	357,469	353,045	351,829	351,078	351,089	342,918	333,177	346,288	352,707	360,351	358,391	362,942
USAR	204,968	206,836	206,892	205,628	206,682	211,890	204,131	189,005	189,975	189,882	197,024	205,297	207,972
USNR	93,171	89,172	86,933	87,913	87,958	88,156	82,558	76,466	70,500	69,933	68,136	66,508	65,493
USMCR	40,842	39,953	39,667	39,810	39,905	41,046	39,644	39,938	39,489	38,557	37,523	38,510	39,459
ANG	108,096	105,715	106,365	108,485	112,071	108,137	106,822	106,430	105,658	106,254	107,679	109,196	108,518
USAFR	71,970	71,772	72,340	73,757	76,632	74,754	75,322	75,802	74,075	71,146	67,565	67,986	69,273
Authorization													
ARNG	361,516	357,223	350,000	350,526	350,000	350,000	350,000	350,000	350,000	350,000	351,300	352,600	358,200
USAR	208,000	208,003	205,000	205,300	205,000	205,000	205,000	205,000	205,000	200,000	205,000	205,000	205,000
USNR	94,294	90,843	90,288	88,900	87,000	87,800	85,900	83,400	73,100	71,300	67,800	66,700	65,500
USMCR	42,000	40,018	39,624	39,558	39,558	39,558	39,600	39,600	39,600	39,600	39,600	39,600	39,600
ANG	108,002	106,992	106,678	108,022	108,400	106,600	107,030	106,800	106,800	107,000	106,700	106,756	106,700
USAFR	73,447	74,243	73,708	74,358	74,700	75,600	75,800	76,100	74,000	74,900	67,500	67,400	69,500
Strength as a percentage of authorization													
ARNG	100	100	101	100	100	100	98	95	99	101	103	102	101
USAR	99	99	101	100	101	103	100	92	93	95	96	100	101
USNR	99	98	96	99	101	100	96	92	96	98	100	100	100
USMCR	97	100	100	101	101	104	100	101	100	97	95	97	100
ANG	100	99	100	100	103	101	100	100	99	99	101	102	102
USAFR	98	97	98	99	103	99	99	100	100	95	100	101	100

Table 2.2—Continued

	1998	1999	2000	2001	2002	2003	2004	2005	2006	2007	2008	2009	2010
Shortage													
ARNG	928	246	3,045	1,303	1,078	1,089	-7,082	-16,823	-3,712	2,707	9,051	5,791	4,742
USAR	-3,032	-1,167	1,892	328	1,682	6,890	-869	-15,995	-15,025	-10,118	-7,976	297	2,972
USNR	-1,123	-1,671	-3,355	-987	958	356	-3,342	-6,934	-2,600	-1,367	336	-192	-7
USMCR	-1,158	-65	43	252	347	1,488	44	338	-111	-1,043	-2,077	-1,090	-141
ANG	94	-1,277	-313	463	3,671	1,537	-208	-370	-1,142	-746	979	2,440	1,818
USAFR	-1,477	-2,471	-1,368	-601	1,932	-846	-478	-298	75	-3,754	65	586	-227

SOURCE: Defense Manpower Data Center.

information on outflow. Because authorizations were constant, the shortfalls in mid-decade were driven by supply factors.[1]

Army Reserve authorization was constant at 205,000 from 2000 to 2010. Major shortfalls occurred in 2004–2008, with shortfalls of 15,000 in 2005 and 2006. As with the Army Guard, these shortfalls were largely supply-driven.

The Navy Reserve shrank by more than a third during the decade, with its authorization declining from 90,000 in 2000 to 65,500 in 2010. Surprisingly, despite this decline, a shortfall was recorded in 2004–2007. We speculate that the shortfall materialized when sailors, realizing that their billet might be eliminated though further reduction in force in coming years, opted to leave sooner than they otherwise would have. The Marine Corps Reserve authorization was constant at 39,600 throughout the decade, but there were shortfalls of 1,000 in 2007 and 2009 and 2,000 in 2008 (or 5 percent of authorization). The Air National Guard and Air Force Reserve had fairly stable authorizations and strength throughout the decade.

Attrition, here defined as leavers as a percentage of average yearly strength, might in principle have been a major contributor to the supply-driven shortfalls experienced by the Army Guard and Reserve and the Marine Corps Reserve. To examine this possibility, we plot in Figure 2.2 Selected Reserve attrition by component for the years available, 2005–2010. (Attrition for 2010 is an annualized estimate based on attrition as of April 2010.)

The attrition trends indicate that Army Guard attrition was steady at around 19 percent, suggesting that its shortfall in 2004–2006 was the product of a decrease in recruits rather than an increase in leavers. Army Reserve attrition was between 20 and 25 percent in 2005–2008 and fell to about 17 percent in 2009–2010. Compared with 2009–2010, the higher rates in 2005–2008 suggest that attrition could have been a factor in the shortfall observed in 2004–2008. This raises the question of what steps were taken to maintain retention. Also, as with the Army Guard, it draws attention to the focus of our research, i.e., how effective RC bonuses were in attracting members exiting from the AC.

Navy Reserve attrition was near or above 30 percent in 2005–2008. This probably reflects higher-than-usual exits caused by downsizing. With the end of downsizing, the attrition rate was just above 20 percent in 2010. Marine Corps Reserve attrition was fairly steady around 24 percent. The shortfalls in 2007–2009 apparently came from too few recruits rather than an increase in leavers. Attrition was lowest and steady at 10 percent in the Air National Guard. Air Force Reserve attrition rose from 15 per-

---

[1] There are no available DMDC data on RC inflow, although there are data on outflow for 2005 to 2010, as shown in Figure 2.2. Using strength and outflow, we have derived inflow for 2005–2009. During these years, there was a net gain of Army National Guard personnel except for 2008, and this contributed to the elimination of all but 1 percent of the Army National Guard shortfall by 2006 and all of it in later years. There were also net gains for the Army Reserve except in 2006, decreases in all years for the Navy Reserve, and mixed changes for the Marine Corps Reserve, Air National Guard, and Air Force Reserve.

cent in 2005 to 19 percent in 2008 and returned to 15 percent in 2010. The Air Force Reserve authorization fell from 74,900 in 2007 to 67,500 in 2008, and the higher attrition in 2007 and 2008 may be a consequence of this cut.

Labor market conditions are generally thought to affect enlistment and retention decisions.[2] The nation entered a deep recession in 2008, with slackening labor market conditions felt as early as 2007. From its low point of 4.5 percent at the beginning of 2007, unemployment increased to 5 percent by the start of 2008 and climbed to nearly 10 percent by mid-2009. High unemployment probably dampens the desire to leave the steady employment offered by the reserves and so decreases attrition. The attrition rates in Figure 2.2 are for all personnel, enlisted and officers, and all ranks and experience levels. The aggregate nature of the data might mask higher unemployment effects among younger workers. In the years before the recession, the unemployment rate fell steadily from just over 6 percent in mid-2003 to 4.5 percent at the end of 2007. The availability of jobs during this period might have encouraged job changing and perhaps movement from one area to another, which could have increased reserve attrition. But there is little evidence of this in the figure.

Figure 2.2
Selected Reserve Attrition, by Component, 2005–2010

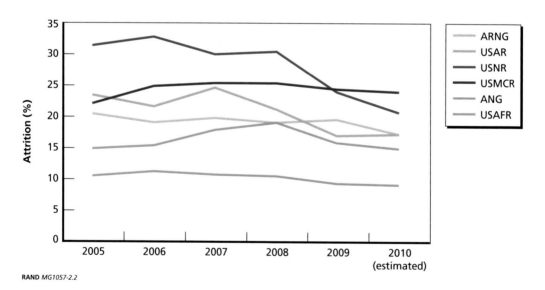

---

[2] For instance, Kostiak and Grogan (1987) find a positive effect of state unemployment rate on state-level recruits into the Navy Reserve, and Hosek and Peterson (1985) find a positive effect of the national unemployment rate on AC reenlistment.

Joining the Selected Reserve from an Active Component

Reserve Bonuses

The Selected Reserve offers two types of "join" bonuses: an affiliation bonus and an enlistment bonus. Prior to 2006, the affiliation bonus was targeted at those who were in an AC with a remaining military service obligation, those who were eligible to reenlist or extend on active duty, those who had 180 days or fewer remaining of active-duty service obligation, and those who left an AC with an honorable discharge and with a remaining military service obligation. The prior service enlistment bonus, or enlistment bonus for short, was targeted at those who completed a period of AC service and were eligible to reenlist. In contrast to the affiliation bonus, the enlistment bonus was not targeted at those with a remaining military service obligation. Also, both types of bonuses are for designated military specialties and grades for which there is a vacancy in the RC.

Affiliation and enlistment bonus eligibility and ceilings have become less restrictive over time, and their domains now intersect. Until 2006, the affiliation bonus was paid at a rate of $50 per month of remaining military service obligation. For instance, the affiliation bonus for 36 months of remaining obligation was $1,800, with a maximum of $6,000. In 2006, the bonus amount was no longer calculated based on the remaining obligation, the ceiling was increased to $10,000 for a three-year reserve enlistment and $20,000 for a six-year reserve enlistment, and eligibility was extended up to 20 years of prior service.

Until 2006, the enlistment bonus was offered only to those who had completed their military service obligation. The window of eligibility was initially narrow; the individual needed to complete the eight-year minimum service obligation but have not more than 10 years of service. From 1999 to 2006, the maximum number of years of prior service gradually increased from 10 to 16 years, serving to expand the eligible population. In 2006, the requirement of no remaining military service obligation was waived, and the ceiling was increased from $5,000 to $7,500 for a three-year reserve contract and from $10,000 to $15,000 for a six-year contract.

To recap, in 2006 the rules governing enlistment and affiliation bonuses became less restrictive and bonus ceilings were raised. These changes expanded the eligible populations and opened the way to increased generosity of bonus awards. Today, affiliation and enlistment bonuses are two similar mechanisms that the RC can use to attract high-quality service members with prior active-duty service. The key difference is that the affiliation bonus has a slightly wider window of eligibility and a slightly higher cap.

Transition from the Active Components to the Selected Reserve

Table 2.3 shows tabulations for AC members at the first- and second-or-higher-term reenlistment points and indicates the percentage choosing to remain in the AC, join

Table 2.3
Transition Rates by Active Component, 1998–2008 Average

Service		Reenlist (%)	Join Selected Reserve (%)	Leave Military (%)
Army	All	43	15	42
	1st term	32	25	43
	2nd term +	49	9	42
Navy	All	47	8	45
	1st term	44	11	45
	2nd term +	48	6	45
Air Force	All	53	8	39
	1st term	51	10	39
	2nd term +	54	6	39
Marine Corps	All	38	5	57
	1st term	27	5	67
	2nd term +	55	4	41

the RC within two years of leaving their AC, or leave the military entirely.[3] Of those facing a reenlistment decision, 15 percent chose to leave the Army and join the reserves, versus 8 percent in the Navy and Air Force and 5 percent in the Marine Corps. The high percentage of Army soldiers at first-term reenlistment who left and joined the reserves is notable: 25 percent.

In both the AC and RC, military operations in Iraq and Afghanistan have required extensive ongoing deployments. Analysis shows that AC enlisted soldiers were heavily stressed, and by 2006 two-thirds of the soldiers at first-term reenlistment had 12 or months of hostile deployment in the previous 36 months (Hosek and Martorell, 2009). For these soldiers and those at second-term reenlistment, the effect of that many months of deployment on reenlistment was negative. The Army's wide use of reenlistment bonuses offset much of the negative effect of deployment, and the Army, like the other services, had fairly stable reenlistment throughout the Operation Iraqi Freedom (OIF)/Operation Enduring Freedom (OEF) period. The Marine Corps also experienced significant deployment and expanded its use of reenlistment bonuses.

The negative effect of extensive AC deployment on Army and Marine Corps reenlistment might have decreased the willingness of prior service soldiers and marines

[3] These tabulations and the following charts are based on our analytical file, which is described in the next chapter. Service members are coded as joining a Selected Reserve component if they join any Reserve component. So, for example, soldiers who join the Navy Reserve are coded as joining the Selected Reserve, even though they switched services.

to join the RC. Our data contain information on AC deployment, and we estimate its effect on RC enlistment.

Figure 2.3 shows AC reenlistment rates (upper panels) and prior service RC enlistment rates (lower panels) by year. The upper panels reveal stability in first- and second-term AC reenlistment rates, though the aggregate rates do not reveal the underlying tug between deployments and bonuses.

The lower panels show that the RC enlistment rates from the AC have not fared as well. Reserve enlistment rates with respect to the first- and second-term-plus reenlistment point show an overall downward trend that began before 9/11. The trend slowed in 2005 and 2006 for the Army and Marine Corps, years when deployment exerted its greatest downward pressure on their AC reenlistment. The slowing may have been caused by the expanded use of bonuses in the RC, shown in Figure 2.4.

With the liberalization of RC affiliation and enlistment bonus authorities in 2006, the Army Guard and Reserve and the Marine Corps Reserve made greater use of these bonuses. Even though the prior service enlistment rates trended down for all the reserve components, we noted that the Army Guard and Reserve and the Marine Corps Reserve had manning shortfalls. Figure 2.4 shows active and reserve bonus usage for AC members at the first-term reenlistment decision point. The patterns for those at second term or higher are similar and not shown. The upper panels of the figure are for AC reenlistment bonuses—percentage receiving and average amount—and the lower panels are for RC enlistment/affiliation bonuses. The RC figures refer to AC members who, having decided not to reenlist, chose to enter the reserves.

In the AC, the percentage of soldiers receiving a reenlistment bonus surged from 2004 to 2005 and increased even further in 2006 and 2007. About 25 percent of soldiers received a reenlistment bonus in 2004, 70 percent in 2005, and over 80 percent in 2007 and 2008. Bonus usage in the Marine Corps increased from 25 percent in 2004 to over 80 percent in 2008. The average amount of the Army and Marine reenlistment bonuses grew by more than 50 percent. By comparison, the percentage of reserve enlistments who received a bonus grew by much less. Among those leaving the Army and entering the reserves, 6 percent received a bonus in 2004, as did 17 percent in 2006. More of those leaving the other services also received a bonus when joining the reserves: 1–2 percent in 2004 versus 5–12 percent in 2008. The average reserve bonus increased from $2,000 or less in 1998–2004 to $8,000–$13,000 in 2008.

Figure 2.3
Active Component Reenlistment and Reserve Join Rate

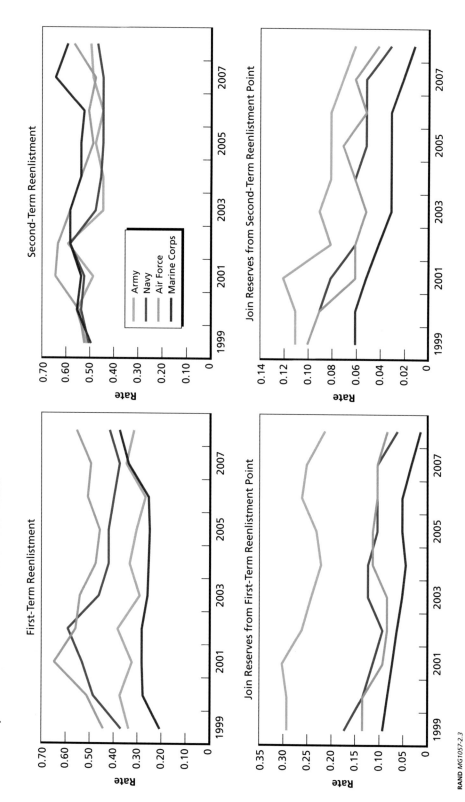

Figure 2.4
Active Component Reenlistment and Reserve Enlistment/Affiliation Bonus Usage

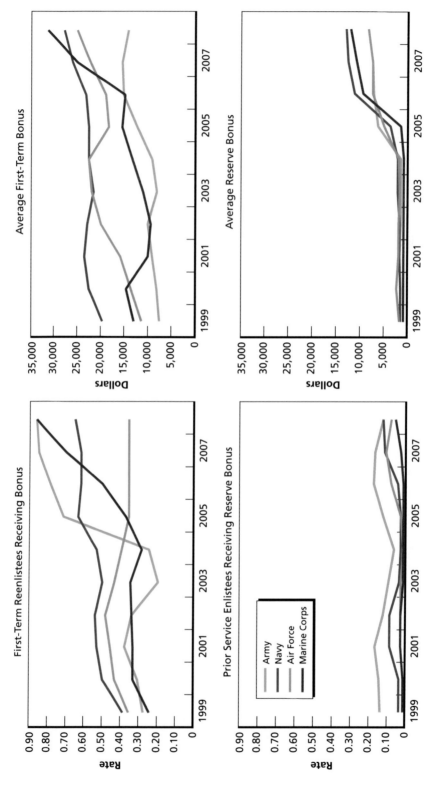

NOTE: Average bonus values are calculated as the mean bonus, conditional on receiving one.

RAND *MG1057-2.4*

CHAPTER THREE

Data and Methods

Data

Our analysis comprises the universe of reenlistment decision points reached by active-duty military personnel with three or more years of service between 1999 and 2008. We link three administrative databases from the Defense Manpower Data Center (DMDC) to create a file with over 2 million individual records containing information about the decision to reenlist, join the Selected Reserve, or leave the military. For each decision point, we have information about bonus payments, military status at the reenlistment point (e.g., service, years of service, pay grade, deployment history), and demographic characteristics (race, gender, and education). We use the information on bonus payments, which are made to service members who reenlist in the AC or join the RC in a specialty offering a bonus, to impute AC and RC bonuses to service members at an AC reenlistment decision point. Appendix A describes the administrative data files and the procedures to generate our decision-level analysis dataset, including the construction of the AC and RC bonus variables, and has tables with the means of the variables used in our analysis file. The tables also include means for the full file.

Construction of the bonus variables required having a sufficient number of observations on individuals who received bonuses to make accurate bonus imputations to all AC members at a reenlistment decision point. Imputation was necessary because there are no data files of bonus offers for all the services. The requirement to have a sufficient number of bonus recipients tended to be met in AC specialties having a high RC join rate, and this resulted in the RC enlistment rate in our analysis file being higher than in the full file. However, the means for all other variables were highly similar between the analysis file and the full file.

Methods

When an AC member's term of service expires, he can reenlist in the AC, join the RC, or become a civilian (C). The AC and RC have the same basic pay table, but the AC is full-time duty and the RC is part-time unless activated. The RC involves weekend drills once a month, individual active-duty training for 14 days in the summer, and possible mobilization and deployment on active duty. RC members typically hold civilian jobs.

We assume that anyone who leaves AC does not reenter it, and anyone who becomes a civilian upon leaving AC does not later, after more than two years, join the RC. These assumptions are realistic. Few ex-AC members rejoin the AC, and most prior AC members who ever join the RC do so within two years after leaving the AC; fewer than 10 percent join more than two years later (Hosek and Martorell, 2009).

We are interested in the effect of RC and AC bonuses on the choice of individual service members to (1) reenlist in the AC, (2) join the RC, or (3) leave the force. Two considerations complicate our analysis. First, our outcome variable has three choice options, so we cannot use standard approaches for binomial choice such as the probit, logit, or linear probability model. Second, we need a framework capable of handling both individual attributes (race, gender, military occupational specialty [MOS], etc.) and choice-based attributes (e.g., the bonus associated with the choice option).

In light of these considerations, we use a version of McFadden's (1974) conditional logit model. The model assumes that individuals have well-defined preferences over a set of choice options and choose the option that is most attractive, conditional on individual- and choice-based attributes. This model has been used in many contexts, including transportation choice (McFadden, 1974), occupational choice (Boskin, 1974), residential choice (Friedman, 1981), college choice (Long, 2004), and college major choice (Montmarquette, Cannings, and Masheredjian, 2002).

The conditional logit is motivated by a random utility model. The individual considers the utility of each alternative, AC, RC, and C. The utilities are assumed to be functions of the bonus, personal characteristics, time and occupational fixed effects, and an error term. The utilities are

$$u_{iA} = a'_1 + \beta_1 B_A + \delta'_1 X_i + \varepsilon_{iA}$$
$$u_{iR} = a'_2 + \beta_2 B_R + \delta'_2 X_i + \varepsilon_{iR}$$
$$u_{iC} = a_3 + \delta_3 X_i + \varepsilon_{iC}.$$

The effects of the AC and RC bonuses, B_A and B_R, and the other variables, X_i, are allowed to differ between the alternatives.[1]

The service member selects the alternative with the highest utility, and this depends in part on the error terms. For instance, the probability of choosing AC is

---

[1]   Note that this model gives a unique set of coefficients δ_i for the effect of individual characteristics on each choice i.

$$\Pr(u_{iA} > u_{iR}, u_{iA} > u_{iC}) = \Pr(\varepsilon_{iA} - \varepsilon_{iR} > a'_2 + \beta_2 B_R + \delta'_2 X - (a'_1 + \beta_1 B_A + \delta'_1 X_i),$$

$$\varepsilon_{iA} - \varepsilon_{iC} > a_3 + \delta_3 X_i - (a'_1 + \beta_1 B_A + \delta'_1 X_i)).$$

Assuming that the error terms are independent draws from a generalized extreme value distribution (GEV), recognizing that the difference of two independent GEV errors is distributed logistically, normalizing with respect to the civilian parameters for identification, and solving for the probabilities of choosing AC and RC (Train, 2009, pp. 78–79), we have expressions for the AC reenlistment and RC enlistment probabilities:

$$r_i = \frac{Exp(a_1 + \beta_1 B_A + \delta_1 X_i)}{1 + Exp(a_1 + \beta_1 B_A + \delta_1 X_i) + Exp(a_2 + \beta_2 B_R + \delta_2 X_i)}$$

$$e_i = \frac{Exp(a_2 + \beta_2 B_R + \delta_2 X_i)}{1 + Exp(a_1 + \beta_1 B_A + \delta_1 X_i) + Exp(a_2 + \beta_2 B_R + \delta_2 X_i)}.$$

The probability expressions imply that the own-bonus effect is positive and the cross-bonus effect is negative.[2]

Using the above expressions for e and r, we estimate the vector of parameters $(\alpha_{hat}, \delta_{hat}, \beta_{hat})$ using the method of maximum likelihood. That is, we solve for the value of $(\alpha_{hat}, \delta_{hat}, \beta_{hat})$ that maximizes the likelihood of observing the choice behavior we observe in the data.

A number of issues make the estimation of bonus effects difficult. These issues include the independence of irrelevant alternatives (IIA) assumption of the logit model, bonus caps, deployment-related bonuses, stop-loss, reverse causality, future bonuses, and the unavailability of vacancies in reserve units. We discuss these issues in Appendix B.

---

[2] From the normalization, $a_1 = a'_1 - a_3$, $a_1 = a'_2 - a_3$, $\delta_1 = \delta'_1 - \delta_3$, and $\delta_2 = \delta'_2 - \delta_3$.

CHAPTER FOUR

Empirical Results and Predictions

A major objective of our research is to determine whether RC bonuses are effective in increasing the enlistment of prior service personnel into the reserves. We are also interested in the interaction between the AC and RC bonuses, and our model allows their effects to differ. This chapter presents our bonus coefficient estimates, projections of AC reenlistment and RC enlistment based on the estimates, and graphics illustrating the bonus effects.

We also present results on two other aspects of RC enlistment: the effect of AC deployment on RC enlistment and the time effects in RC enlistment. The prevalence of deployment in the AC grew rapidly because of the operations in Iraq and Afghanistan, and by 2004 the majority of soldiers and marines at reenlistment had been deployed. For the RC, the question is whether extensive AC deployment affected RC enlistment. The time effects control for other covariates, which include AC and RC bonuses, years of service, education, Armed Forces Qualification Test (AFQT) category, one-digit occupation, months of deployment in the AC in the three years before the reenlistment decision point, whether the service member was promoted faster than average, gender, race/ethnicity, and marital status. After controlling for these variables, the time effects reveal whether there has been a trend in RC enlistment.

In brief, we find that AC and RC bonuses are effective, but their effects differ. The RC bonus effect estimates are quadratic and positive at lower bonus values. The RC bonus effect reaches its maximum at bonus levels of $10,000–$13,000, though as suggested in Appendix B this tapering off may result from bonus ceilings. The AC bonus effect estimates are also quadratic, but the effect is only mildly curved over the range where most AC bonuses are paid, and its maximum occurs at a much higher bonus value.

We find that AC deployment tends to have a small effect on RC enlistment. Exceptions are the Army, where deployment of less than nine months increases first- and second-term reenlistment and decreases RC enlistment, and the Navy, where hostile AC deployment increases second-term reenlistment and decreases RC enlistment.

Finally, we find negative time trends for RC enlistment. Notably, the trends began before 9/11 and have continued since then. The reasons for the trends are unknown, but we speculate that the downtrend began because of the economic boom at the end of the 1990s and was influenced after 9/11 by the growing likelihood that joining the RC would lead to deployment. For individuals who preferred not to deploy, perhaps because they were focused on their civilian work and personal lives, a higher expectation of deployment might have deterred their enlistment in the RC. Other factors could have been the cause, however. For instance, perhaps RC prior service recruiting became less effective relative to AC reenlistment efforts during this period.

Bonus Coefficient Estimates

Table 4.1 contains the bonus coefficients and their standard errors for AC and RC bonuses. We find strong, statistically significant, positive coefficients for both bonuses. The quadratic terms are significant in all cases except the Marine Corps Reserve, and the general pattern is that bonuses increase the probability of reenlistment, or enlistment into the reserves, and the rate of increase tapers off as the size of the bonus increases. Although this occurs in the AC, it is more the case in the RC where the bonus effects reach a maximum at bonuses in the range of $10,000–$13,000.[1]

We tested whether bonus effects differ in combat arms specialties versus non–combat arms specialties but found only marginal evidence of differences. We also tested for bonus interactions with deployment history and found little evidence to support the idea that bonus effects vary by past deployment.

Projections of Bonus Effects

Using our estimates, we made projections of AC reenlistment rates and RC enlistment rates by bonus amounts in each component. Tables 4.2–4.5 report projections for first-term service members by service for an AC bonus range of $0–$16,000 and an RC range of $0–$8,000. The tables show predicted AC and RC probabilities at different bonus levels, with other variables held constant at their mean.[2] We do not report standard errors or confidence intervals for these projections. Given the large sample sizes, most differences are statistically significant. We have computed 95 percent confidence intervals for several values, and in general, they range in size from 1–2 percentage points around the point estimates. The formulas for the probabilities are those presented in Chapter Three.

---

[1] The maximum occurs at a point determined by the negative of the linear coefficient divided by twice the quadratic coefficient. For example, the Navy AC bonus effect reaches its maximum at $0.0639/(2 \times .0012) = \$26,600$ and the Navy RC bonus effect reaches its maximum at $0.0458/(2 \times 0.0018) = \$12,700$.

[2] AC reenlistment rates and RC enlistment rates are defined as the fraction of service members at a reenlistment decision point who reenlist in the AC (join the RC within two years), regardless of the service in which they enlist. Thus, for example, if an Army soldier were to join the Navy Reserves, that would add to the Army RC join rate.

Table 4.1
Bonus Coefficients and Standard Errors, by Service and Term

	Army	Navy	Marine Corps	Air Force
First term				
AC bonus	.0344***	.0639***	.0899***	.0155***
AC bonus squared	−.0003*	−.0012***	−.0017***	−.0002**
RC bonus	.0370***	.0458***	.0448	.0439***
RC bonus squared	−.0019***	−.0018***	−.0012	−.0017***
	SE	SE	SE	SE
AC bonus	.0030	.0024	.0037	.0024
AC bonus squared	.0001	.0001	.0001	.0001
RC bonus	.0033	.0082	.0332	.0084
RC bonus squared	.0002	.0004	.0023	.0005
Second+ term				
AC bonus	.0540***	.0463***	.0534***	.0322***
AC bonus squared	−.0009***	−.0007***	−.0009*	−.0009***
RC bonus	.0178***	.0656***	−.0413	.0193
RC bonus squared	−.0009**	−.0031***	−.0001	−.0015*
	SE	SE	SE	SE
AC bonus	.0028	.0035	.0098	.0034
AC bonus squared	.0001	.0001	.0004	.0001
RC bonus	.0050	.0127	.0955	.0132
RC bonus squared	.0003	.0007	.0059	.0008

NOTES: The table reports the bonus coefficients (the b coefficients in our conditional logit model) and their standard errors for the active and reserve bonuses. We allow the AC and RC bonuses to have a quadratic relationship with the utility of choosing AC and RC. SE is the standard error of the coefficient. Statistical significance: *** = 0.001, ** = 0.01, * = 0.1.

In general, predicted AC reenlistment rates increase with the AC bonus and decrease with the RC bonus, while RC enlistment rates increase with the RC bonus and decrease with the AC bonus. For instance, in Table 4.2 when the reserve bonus is $2,000, increasing the Army reenlistment bonus from $4,000 to $8,000 increases its reenlistment rate from 0.378 to 0.408. But it also decreases the reserve enlistment rate from 0.248 to 0.236. At the new AC bonus of $8,000, the reserve bonus must be increased to an amount somewhat above $4,000 to hold reserve enlistment constant. We consider this further in Chapter Five in discussing bonus costs.

Table 4.2
Projected AC Reenlistment and RC Join Rates Among First-Term Army Personnel, by AC Reenlistment Bonus and RC Enlistment/Affiliation Bonus

AC Reenl. Bonus	Selected Reserve Enlistment/Affiliation Bonus									
	$0		$2,000		$4,000		$6,000		$8,000	
	Reenlist in AC	Join RC	Reenlist in AC	Join RC	Reenlist in AC	Join RC	Reenlist in AC	Join RC	Reenlist in AC	Join RC
$0	.354	.247	.348	.260	.343	.270	.340	.277	.338	.281
$4,000	.385	.235	.378	.248	.374	.257	.370	.264	.368	.268
$8,000	.415	.224	.408	.236	.403	.245	.400	.252	.398	.256
$12,000	.443	.213	.437	.224	.432	.233	.428	.240	.426	.244
$16,000	.470	.203	.464	.214	.459	.222	.455	.229	.453	.232

Table 4.3
Projected AC Reenlistment and RC Join Rates Among First-Term Navy Personnel, by AC Reenlistment Bonus and RC Enlistment/Affiliation Bonus

AC Reenl. Bonus	Selected Reserve Enlistment/Affiliation Bonus									
	$0		$2,000		$4,000		$6,000		$8,000	
	Reenlist in AC	Join RC	Reenlist in AC	Join RC	Reenlist in AC	Join RC	Reenlist in AC	Join RC	Reenlist in AC	Join RC
$0	.282	.177	.278	.190	.274	.201	.271	.210	.269	.217
$4,000	.333	.165	.328	.177	.324	.187	.320	.196	.317	.203
$8,000	.378	.154	.373	.165	.368	.175	.365	.183	.362	.189
$12,000	.416	.144	.410	.155	.406	.164	.402	.172	.399	.178
$16,000	.445	.137	.440	.147	.435	.156	.431	.164	.428	.170

Table 4.4
Projected AC Reenlistment and RC Join Rates Among First-Term Marines, by AC Reenlistment Bonus and RC Enlistment/Affiliation Bonus

| AC Reenl. Bonus | Selected Reserve Enlistment/Affiliation Bonus | | | | | | | | | |
| | $0 | | $2,000 | | $4,000 | | $6,000 | | $8,000 | |
	Reenlist in AC	Join RC	Reenlist in AC	Join RC	Reenlist in AC	Join RC	Reenlist in AC	Join RC	Reenlist in AC	Join RC
$0	.189	.105	.187	.113	.185	.121	.184	.128	.182	.135
$4,000	.245	.098	.243	.106	.241	.113	.239	.120	.237	.126
$8,000	.299	.091	.297	.098	.295	.105	.293	.111	.291	.117
$12,000	.348	.085	.345	.091	.343	.098	.340	.104	.338	.109
$16,000	.386	.080	.383	.086	.381	.092	.378	.098	.376	.103

Table 4.5
Projected AC Reenlistment and RC Join Rates Among First-Term Air Force Personnel, by AC Reenlistment Bonus and RC Enlistment/Affiliation Bonus

| AC Reenl. Bonus | Selected Reserve Enlistment/Affiliation Bonus | | | | | | | | | |
| | $0 | | $2,000 | | $4,000 | | $6,000 | | $8,000 | |
	Reenlist in AC	Join RC	Reenlist in AC	Join RC	Reenlist in AC	Join RC	Reenlist in AC	Join RC	Reenlist in AC	Join RC
$0	.490	.168	.483	.180	.477	.190	.472	.198	.468	.205
$4,000	.504	.163	.497	.175	.492	.185	.487	.193	.483	.199
$8,000	.517	.159	.510	.170	.504	.180	.500	.188	.496	.194
$12,000	.529	.155	.522	.166	.516	.176	.511	.184	.507	.190
$16,000	.538	.152	.531	.163	.525	.172	.520	.180	.517	.186

In all cases, a bonus increase in either component increases the sum of the AC reenlistment rate and the RC enlistment rate. That is, a higher fraction stays in the military. We find similar results for second-term personnel and report them in Appendix D.

Deployment Effects on AC Reenlistment and RC Enlistment

Table 4.6 shows coefficients for nonhostile and hostile AC deployment with respect to AC reenlistment and RC enlistment.[3] We use the coefficients to compute the effect of deployment on the probabilities of AC reenlistment and RC enlistment (see Figures 4.1–4.4). We also decompose the effect of AC deployment on RC enlistment into parts resulting from the change in AC leavers (non-reenlistment) and the change in RC enlistment given leaving AC. The decomposition isolates the mechanical effect of deployment on RC enlistment caused by the change in the number of AC leavers from its effect on RC enlistment among AC leavers.

In the table, the coefficient for hostile deployment indicates its difference from the nonhostile deployment effect, and the nonhostile and hostile coefficients add together to yield the overall effect of hostile deployment. Consistent with previous work (Hosek and Martorell, 2009), longer deployment has a negative effect on first-term reenlistment for the Army and Marine Corps.

The effects of nonhostile deployment on AC first-term reenlistment (Figure 4.1) are large for the Navy and Air Force: 8.2 and 8.8 percentage points for short deployment and 10.7 percentage points, and 13.7 percentage points for long deployment, respectively. The Army effects are also positive though smaller, and the Marine Corps effects are near zero or slightly negative. In contrast, the effects on RC enlistment are small in all cases. The effects on second-term AC reenlistment (Figure 4.2) are large, ranging from 6 to 18 percentage points for the Army, Navy, and Marine Corps, but are negative and statistically insignificant for RC enlistment.

Figures 4.3 and 4.4 show the effects of hostile deployment. First-term Army personnel respond much differently to long versus short hostile deployment (Figure 4.3). We estimate that short deployment increases the reenlistment of first-term soldiers by 8.0 percentage points and decreases the probability of joining the RC by 4.9 percentage points. In contrast, our results suggest that long deployment *decreases* reenlistment by 7.5 percentage points but has little effect on RC enlistment. Overall, the results indicate little overall effect of first-term AC hostile deployment on RC enlistment, with the exception of Army short hostile deployment, which increases AC reenlistment and decreases enlistment in the RC.

---

[3]   To be consistent with Hosek and Martorell (2009), we allow the effect of deployment to vary by whether a service member was deployed 0 months, 1–8 months, or 9 or more months.

Table 4.6
Deployment Coefficients and Standard Errors, by Service and Term

Deployment	Army AC	Army RC	Navy AC	Navy RC	Marine Corps AC	Marine Corps RC	Air Force AC	Air Force RC
First term								
Deployed 1–8 months	.2072	–.0398	.3808	.1134	.0086	.0985	.4327	.2367
	.0270	.0292	.0368	.0509	.0382	.0535	.0375	.0517
Deployed 9+ months	.1899	.1516	.4963	.1872	–.1170	.1858	.6548	.2948
	.0325	.0344	.0435	.0599	.0458	.0632	.0546	.0741
Hostile deployed 1–8 months	.1931	.0668	–.2566	–.0655	.0170	–.0527	–.3338	.0130
	.0269	.0293	.0340	.0479	.0364	.0511	.0374	.0514
Hostile deployed 9+ months	–.5019	.0247	–.3657	–.0650	–.0951	–.2182	–.6122	.1083
	.0343	.0358	.0521	.0705	.0509	.0724	.0603	.0798
Second term								
Deployed 1–8 months	.3527	.1233	.3901	.0309	.5541	.0315	.2269	.2322
	.0265	.0345	.0448	.0569	.0814	.1186	.0449	.0576
Deployed 9+ months	.4554	.2357	.7610	.1086	.4620	.0367	.3654	.2522
	.0289	.0377	.0633	.0823	.1215	.1799	.0584	.0754
Hostile deployed 1–8 months	.1511	–.0337	–.0359	–.0328	–.0172	.1650	.0812	.0290
	.0264	.0350	.0455	.0585	.0937	.1372	.0444	.0568
Hostile deployed 9+ months	–.2941	–.0221	–.0751	–.0434	–.1013	.1057	.0597	.1834
	.0318	.0429	.0985	.1332	.1680	.2504	.0735	.0939

Figure 4.1
Effect of Nonhostile Deployment on First-Term AC Reenlistment Rates and RC Enlistment Rates, by Service

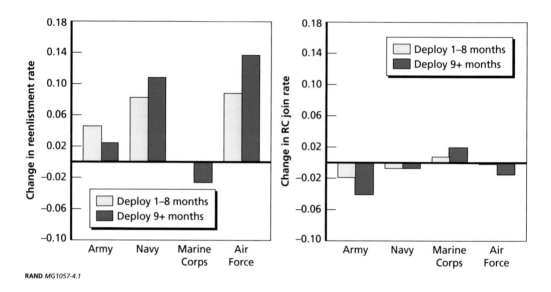

RAND *MG1057-4.1*

Figure 4.2
Effect of Nonhostile Deployment on Second-Term AC Reenlistment Rates and RC Enlistment Rates, by Service

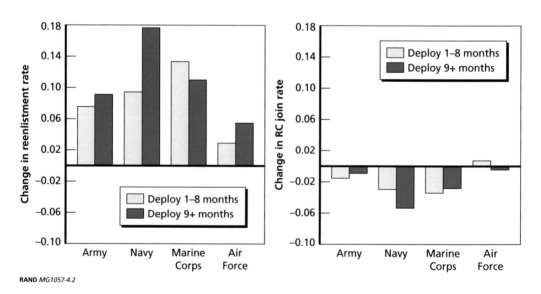

RAND *MG1057-4.2*

Figure 4.3
Effect of Hostile Deployment on First-Term AC Reenlistment Rates and RC Enlistment Rates, by Service

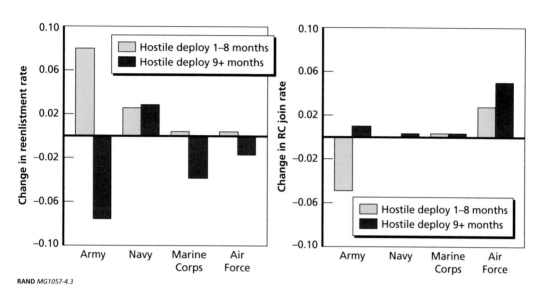

RAND *MG1057-4.3*

Figure 4.4
Effect of Hostile Deployment on Second-Term AC Reenlistment Rates and RC Enlistment Rates, by Service

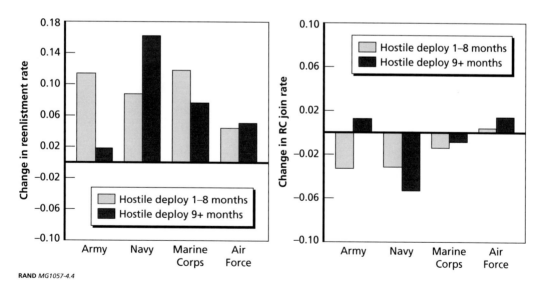

RAND *MG1057-4.4*

The Army and Marine Corps have been heavily involved in ground combat operations in Iraq and Afghanistan, and it is somewhat surprising that hostile deployment of 1–8 months increases Army first-term reenlistment but has no effect on Marine Corps first-term reenlistment. Our deployment coefficients are the average effect across the years of our study (1999–2008). A positive average effect of 1–9 month of hostile deployment on Army first-term reenlistment and a zero effect on Marine first-term reenlistment correspond well with the findings of Hosek and Martorell (2009). As to why the effects should differ between the Army and Marine Corps, one possibility is deployment length. Army deployments to Iraq and Afghanistan have been 12–15 months long, as compared with Marine Corps deployments that have been 7 months. Many soldiers with 1–9 months of hostile deployment probably did not experience that deployment in Iraq or Afghanistan, whereas many of the marines with 1–9 months of hostile deployment probably did.

Our estimates suggest that the effects of hostile deployment on second-term reenlistment are positive and sometimes large (Figure 4.4). We find that short hostile deployment boosts Army reenlistment by nearly 12 percentage points, while long deployment has a near-zero effect. Our estimates for short- and long-deployment effects on Navy and Marine Corps reenlistment are above 6 points and range higher. Finally, with respect to RC enlistment, our estimates suggest that hostile deployments have a zero or negative impact.

When we decompose our estimates for the effect of AC deployment on RC enlistment, we find that most of the effect comes from the change in the number of AC leavers, and little comes from the change in RC enlistment among AC leavers. The formula for the decomposition has a marginal effect and a conditional effect:

$$p(RC\ enlistment) = p(leave\ AC)\ p(enlist\ in\ RC|leave\ AC).$$

The first term on the right-hand side is the marginal effect, i.e., the effect of AC deployment on leaving AC and therefore being at risk to join the RC, and the second term is the conditional effect on RC enlistment given leaving AC.

We use this formula to see how a change in a variable, e.g., the occurrence of deployment, changes the probability of RC enlistment:

$$\Delta\ p(RC\ enlistment) = \Delta\ p(leave\ AC)\ p(enlist\ in\ RC|leave\ AC) + \\ \Delta\ p(enlist\ in\ RC|leave\ AC)\ p(leave\ AC).$$

Table 4.7 shows computed values of the decomposition based on our estimated coefficients and sample mean probabilities.

In nearly all cases, AC deployment decreases the probability of leaving AC, which decreases the pool available to join the RC. This makes the term for the marginal effect in the formula negative. However, AC deployment increases the probability of RC enlistment conditional on leaving AC, such that the term for the conditional effect

Table 4.7
Decomposition of the Effect of AC Deployment on the Probability of RC Enlistment

	From Δ in Leaving AC	From Δ in RC Enlistment Given Leave AC	Total Δ in RC Enlistment	From Δ in Leaving AC	From Δ in RC Enlistment Given Leave AC	Total Δ in RC Enlistment
Non-hostile		1–8 months			9+ months	
First term						
Army	−.020	−.007	−.027	−.011	.027	.016
Navy	−.022	.015	−.007	−.028	.024	−.004
Marine Corps	.000	.008	.008	.004	.016	.020
Air Force	−.029	.027	−.002	−.046	.031	−.015
Second term						
Army	−.027	.012	−−.015	−.032	.023	−.009
Navy	−.034	.001	−.033	−.057	.010	−.047
Marine Corps	−.037	.002	−.035	−.028	.002	−.026
Air Force	−.012	.008	−.004	−.024	.020	−.004
Hostile		1–8 months			9+ months	
First term						
Army	−.036	.004	−.032	.034	.029	.063
Navy	−.007	.005	−.002	−.007	.010	.003
Marine Corps	.000	.004	.004	.005	−.003	.002
Air Force	−.002	.029	.027	.006	.040	.046
Second term						
Army	−.041	.008	−.033	−.007	.019	.012
Navy	−.030	.002	−.028	−.054	.019	−.035
Marine Corps	−.032	.019	−.013	−.021	.013	−.008
Air Force	−.019	.033	.014	−.021	.036	.015

in the formula is positive. The marginal effect often dominates the conditional effect, resulting on net in a negative effect of AC deployment on RC enlistment. An exception occurs for first-term hostile deployment of 9 or more months, where the net effect of deployment on RC enlistment is positive for the Army and Air Force though nearly zero for the Navy and Marine Corps.

Year Fixed Effects Reveal Downward Trends in RC Enlistment Rates

Figure 4.5 presents the year effects on RC enlistment for first-term service members. The estimates are relative to the base year of 2008.[4] The year effects control for bonuses, past deployment, and other variables in the model, whereas the simple trends shown in Chapter Two do not control for these factors. We were surprised to find a clear downward trend in RC enlistment for each service.

The cumulative impact of the trend is noticeable: RC enlistment fell by approximately 7 percentage points from 2001 to 2008, even after controlling for the other factors in the model. In the Army, Navy, and Marine Corps, the trend began before 9/11 and may have been influenced by the economic boom.[5] Also, as deployments have become more frequent after 9/11, personnel leaving the AC might have been concerned about being deployed if they joined the RC. Perhaps they did not want to be deployed, having been deployed in the AC, or were concerned that deployment would disrupt their civilian career and family life. Further, we do not know whether RC prior service recruitment efforts weakened during the decade, or weakened relative to AC reenlistment efforts. The downward trend might also reflect demand, especially for the Navy Reserve, which was downsizing (see Table 2.2). But when we calculate the annual inflow to the RC for 2005–2009 (as noted in Chapter Two), Navy Reserve

Figure 4.5
First-Term Year Effects on RC Enlistment, by Service

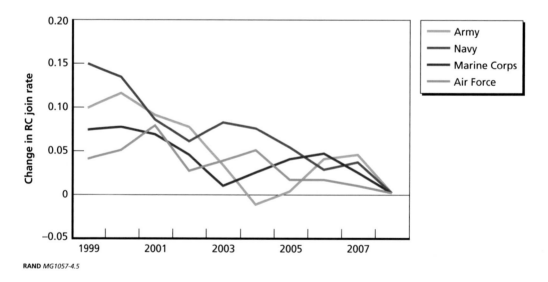

RAND MG1057-4.5

---

[4] For 2008, the final year in our data, there was not enough time to follow all service members who left active duty into the RC for the full two-year follow-up period. This biases the coefficient on the 2008 dummy toward zero but does not affect the finding that there is a downward trend in year effects over time in earlier years.

[5] We discuss this further in Chapter Six.

inflows were fair stable through 2008. This suggests that downsizing does not account for the downward trend in the Navy Reserve.[6] Other trend data, i.e., ad-tracking data and youth poll surveys collected by the Joint Advertising and Marketing Research System (JAMRS), show that youth and young adults have a negative attitude toward the war on terrorism and that youth propensity to enlist declined between 2001 and 2007 (Asch et al., 2010). The JAMRS data are for the youth and young adult population, while the trends in Figure 4.5 come from individuals who served in the AC and left after the first term. Yet perhaps their attitudes toward further involvement in OIF/OEF were similar to those in the youth population.

---

[6] The inflows were 17,900 in 2005, 22,400 in 2006, 19,100 in 2007, 19,000 in 2008, and 14,800 in 2009.

CHAPTER FIVE

Bonus Setting and Bonus Cost in the Active and Reserve Components

---

The choice model in Chapter Three includes AC and RC bonuses but does not consider how bonuses are set. In this chapter, we examine this question by applying ideas from economic theory. We focus on two alternative objectives of the bonus setter, described below. Under either objective, the RC bonus setter responds to a higher AC bonus with a higher RC bonus, and vice versa.

In keeping with Tables 4.2–4.5 in Chapter Four, we compute bonus costs for a range of AC and RC bonus amounts. However, treating bonus amounts as the driver of AC reenlistment and RC enlistment has a limitation. It does not identify the bonus, and bonus cost, that one component must pay to hold its rate constant when the other component increases its bonus. To overcome this limitation, we also compute bonuses and bonus costs in a way that accounts for these externalities.

Bonus Setting

The bonus-setting objectives we consider are (1) targets for AC reenlistments and RC enlistments and (2) maximizing the value of additional AC reenlistments and RC enlistments relative to their bonus costs. The optimal bonuses for the first objective depend solely on the targets. For the second objective, they depend on the interaction of the AC and RC bonus effects. Under either objective, an increase in one component's bonus creates a negative externality for the other component as a result of the supply-side response; to offset this, the other component must increase its bonus. The need for this response derives from the individual's decisionmaking behavior in which changing either bonus affects both the AC reenlistment and RC enlistment probabilities. It does not reflect any inefficiency in bonus setting leading to unnecessarily high bonuses. Coordination between AC and RC bonus setters will not eliminate this cross-effect of bonuses but can prevent surprises and inadequate budgeting.

AC and RC Target Rates

Suppose the AC and RC targets are R reenlistments and E enlistments, respectively. With the number of service members at AC reenlistment n given, the targets can be stated in terms of rates $R/n = r^T$ and $E/n = e^T$. In the supply model, each rate depends on the AC and RC bonuses: $r = r(B_A, B_R)$ and $e = e(B_A, B_R)$, and the bonus setters find their optimal bonuses by each solving these relationships for their bonus as a function of the rates, which are interpreted as given targets.[1]

In our empirical specification, the bonuses enter quadratically, and the AC and RC bonus parameters differ from one another. The probability functions may be abbreviated as

$$r = \frac{Exp(c_1 + \beta_1 B_A + \gamma_1 B_A^2)}{1 + Exp(c_1 + \beta_1 B_A + \gamma_1 B_A^2) + Exp(c_2 + \beta_2 B_R + \gamma_2 B_R^2)}$$

$$e = \frac{Exp(c_2 + \beta_2 B_R + \gamma_2 B_R^2)}{1 + Exp(c_1 + \beta_1 B_A + \gamma_1 B_A^2) + Exp(c_2 + \beta_2 B_R + \gamma_2 B_R^2)}.$$

Here, the constants of c_1 and c_2 represent all the non-bonus variables evaluated at some point. Solving for the bonuses with respect to targets r and e, the optimal bonuses are

$$B_A^* = \frac{-\beta_1 + \sqrt{\beta_1^2 - 4c_1\gamma_1 + 4\gamma_1 Log\left(\frac{r}{1-r-e}\right)}}{2\gamma_1}$$

$$B_R^* = \frac{-\beta_2 + \sqrt{\beta_2^2 - 4c_2\gamma_2 + 4\gamma_2 Log\left(\frac{e}{1-r-e}\right)}}{2\gamma_2}.$$

These formulas can be used to find bonuses that correspond to given AC reenlistment and RC enlistment rates. For instance, suppose that at some initial bonus levels the rates were (r_0, e_0), and the AC wanted to increase its rate to $r_1 > r_0$ while the RC wanted to keep its rate constant. The new rates (r_1, e_0) can be entered into the formulas to find the new AC and RC bonuses. Just as Tables 4.2–4.5 showed the reenlistment and enlistment rates corresponding to each AC and RC bonus combination, these formulas can be used to compute the bonuses corresponding to each reenlistment and

---

[1]   One can imagine that a bonus setter for a component first inverts its supply equation to obtain its bonus as a function of its target rate and the bonus of the other component. The bonus setter for the other component will do the same. These may be thought of as best-response functions. Given the bonus of the other component, the bonus setter will adjust its bonus in accord with its rate. But as this bonus changes, the other bonus setter must also change its bonus. This process reaches an equilibrium when the AC and RC bonuses are simultaneously consistent with the AC and RC target rates.

enlistment rate combination. We discuss this idea further when presenting bonus cost estimates later in the chapter.

To make use of the formulas, we first apply our estimated bonus parameters to find values of c_1 and c_2 that reflect the mean reenlistment and enlistment rates in the absence of the bonus.[2] We can then, for example, ask what AC and RC bonuses support Army first-term targets of $r = 0.42$ and $e = 0.24$. The answer is an AC bonus of $10,517 and an RC bonus of $4,648.

The formulas imply that each of the optimal bonuses is higher the higher each of the target rates. In effect, if the own-target rate increases, the bonus setter must move along the supply curve, which slopes upward, and pay a higher bonus in order to meet the target. If the other bonus increases, it has a negative cross-effect on the own rate, and the bonus setter offsets this by increasing its own bonus.

Value Net of Bonus Cost

This case assumes notional functions for the values of additional AC and RC members. First, suppose that AC and RC budget setters act independently to maximize their value subject to a supply function, given the bonus set by the other. Each additional member is valuable but comes at a higher bonus cost:[3]

$$V_A(A) - AB_A$$
$$V_R(R) - RB_R.$$

Here, V_A is the AC value function, A is the number of reenlistments ($A = nr$), AB_A is the bonus cost, and similarly for the RC. The first-order conditions equate the marginal value of the bonus to its marginal cost:

---

[2]   The values of c_1 and c_2 are obtained by solving the probability functions to obtain

$$c_1 = -\beta_1 B_A - \gamma_1 B_A^2 + Log(r / (1 - r - e))$$
$$c_2 = -\beta_2 B_R - \gamma_2 B_R^2 + Log(e / (1 - r - e)).$$

We evaluate these expressions at the mean bonuses and the mean reenlistment and enlistment rates. Thus, controlling for the bonuses, the calculated values of c_1 and c_2 are those consistent with the mean reenlistment and enlistment rates.

[3]   Bonuses may create secondary effects. For instance, by increasing reenlistment the AC bonus increases the number of AC members who will reach future AC reenlistment points, in turn increasing the potential number of prior service enlistments into RC. The secondary effects are not included. Also, military pay, such as basic pay and allowances, can be included in the cost, but the basic analysis remains the same.

$$n\frac{\partial V_A}{\partial A}\frac{\partial r}{\partial B_A} = n\left(r + \frac{\partial r}{\partial B_A}B_A\right)$$

$$n\frac{\partial V_R}{\partial R}\frac{\partial e}{\partial B_R} = n\left(e + \frac{\partial e}{\partial B_R}B_R\right).$$

These conditions implicitly define the optimal bonus amounts. The conditions use the fact that $\partial A / \partial r = n$ and $\partial R / \partial e = n$. The marginal value reflects the incremental value of an increase in AC reenlistments or RC prior service enlistments, respectively, resulting from an incremental increase in the AC or RC bonus. The marginal cost reflects paying an incrementally higher bonus to those who were willing to sign up at the initial level of the bonus (the nr and ne terms), plus paying the full bonus to the increase in the number willing to sign up (the increase in probability terms).

The conditions can be expressed in terms of the marginal cost per additional reenlistment or enlistment by dividing through by $\partial r / \partial B_A$ and $\partial e / \partial B_R$, respectively:

$$\frac{\partial V_A}{\partial A} = \frac{r}{\dfrac{\partial r}{\partial B_A}} + B_A$$

$$\frac{\partial V_R}{\partial R} = \frac{e}{\dfrac{\partial e}{\partial B_R}} + B_R.$$

Each equation is a best-response curve giving the optimal bonus as a function of the own bonus and other bonus. An equilibrium is achieved when the values of B_A^* and B_R^* satisfy both equations simultaneously. The equations cannot be solved for explicit or numerical solutions of the optimal bonus values without knowing the value functions, but it is possible to describe the cross-effects of bonus changes. In particular, a higher RC bonus results in a higher optimal AC bonus. This is because the higher RC bonus causes the reenlistment rate to decrease, so the first term on the right-hand side of the AC equation decreases. To restore equality, the AC bonus must increase. By the same reasoning, a higher AC bonus results in a higher optimal RC bonus.

We can place these conditions in the context of the multinomial logistic probability model. In that model, the derivatives of the AC and RC probabilities with respect to own bonus are $\partial r / \partial B_A = \beta r(1-r)$ and $\partial e / \partial B_R = \beta e(1-e)$. Using these expressions,

$$\frac{\partial V_A}{\partial A} = \frac{1}{\beta(1-r)} + B_A$$
$$\frac{\partial V_R}{\partial R} = \frac{1}{\beta(1-e)} + B_R.$$

As before, the left-hand side is the marginal value and the right-hand side is the marginal cost.[4]

With empirical estimates of the probability parameters and the bonuses, one can evaluate the right-hand side and obtain estimates of the marginal costs. They also represent point estimates of the marginal values in the context of this model.

Joint Bonus Setting by AC and RC

We extend the analysis to consider joint bonus setting by the AC and RC. Changing a bonus creates a cross-effect, and joint bonus setting internalizes the cross-effect. If the AC bonus is increased, the RC enlistment rate will decrease unless the RC bonus is increased, and vice versa. The amount by which the other bonus must be adjusted can be thought of as a part of the full cost, or alternatively as a compensating variation in expenditure.

The joint function is

$$V_A(A) + V_R(R) - AB_A - RB_R.$$

The first-order conditions per additional AC reenlistment and RC enlistment again equate marginal value and marginal bonus cost:

$$\frac{\partial V_A}{\partial A} = B_A + \frac{r}{\dfrac{\partial r}{\partial B_A}} - \frac{\dfrac{\partial e}{\partial B_A}}{\dfrac{\partial r}{\partial B_A}} B_R$$

$$\frac{\partial V_R}{\partial R} = B_R + \frac{e}{\dfrac{\partial e}{\partial B_R}} - \frac{\dfrac{\partial r}{\partial B_R}}{\dfrac{\partial e}{\partial B_R}} B_A.$$

---

[4] In our quadratic bonus specification, β in the equations in the text above is replaced by $\beta_1 + 2\gamma_1 B_A$ in the AC equation and $\beta_2 + 2\gamma_2 B_A$ in the RC equation. Our estimates find that the betas are positive and the gammas are negative, which implies that the quadratic initially slopes upward. Further, as the bonus effect approaches its maximum, which occurs where $\beta + 2\gamma B = 0$, these expressions approach zero. This causes the marginal cost to approach infinity.

These are the same first-order conditions as before but with the addition of the cross-effect terms.[5] The partial derivatives $\partial e / \partial B_A$ and $\partial r / \partial B_R$ are negative, so the third terms on the right-hand side, which enter with a minus sign, are positive. They represent the additional bonus cost imposed on the other component by increasing the own bonus. The other bonus is increased to hold its rate at its initial value. The ratio of partial derivatives in the third term expresses the rate at which an RC enlistment can be "traded" for an AC reenlistment at the equilibrium.

Bonus Cost

We calculate bonus costs from two perspectives. The first gives AC and RC bonus costs for different combinations of AC and RC bonus amounts, and the second gives AC and RC bonus costs for different combinations of AC reenlistment and RC enlistment rates. The first perspective corresponds to Tables 4.2–4.5, which display reenlistment and enlistment rates for various AC and RC bonus amounts. These tables have an intuitive appeal because they show how the rates change as the bonuses change, but they also have a limitation. In particular, increasing either bonus while holding the other bonus constant changes *both* rates. For example, an increase in the AC bonus while holding the RC bonus constant results in an increase in the AC reenlistment rate and a decrease in the RC enlistment rate. The second perspective overcomes this limitation. It presents bonus costs as the AC reenlistment rate increases and the RC enlistment rate is held constant, and vice versa. This perspective shows the bonus sizes and bonus costs needed to sustain each AC-RC rate combination. In this case, the bonus costs include the own bonus cost and, separately, the other bonus cost incurred to hold the other rate constant. This indirect cost, or externality, is the cost counterpart to the cross-effect marginal cost discussed above.

---

[5]   In our quadratic specification, the marginal costs are

$$B_A + \frac{1}{(\beta_1 + 2\gamma_1 B_A)(1-r)} + \frac{Exp(c_2 + \beta_2 B_R + \gamma_2 B_R^2)}{1 + Exp(c_2 + \beta_2 B_R + \gamma_2 B_R^2)} B_R$$

$$B_R + \frac{1}{(\beta_2 + 2\gamma_2 B_R)(1-e)} + \frac{Exp(c_1 + \beta_1 B_A + \gamma_1 B_A^2)}{1 + Exp(c_1 + \beta_1 B_A + \gamma_1 B_A^2)} B_A.$$

With parameter estimates, values of c_1 and c_2 (see footnote above), and bonus values, these formulas can be used to calculate marginal cost at a point of evaluation designated by the reenlistment and enlistment rates and the bonus amounts.

Bonus Cost with Respect to AC and RC Bonus Amounts

The joint cost function in terms of bonus amounts is

$$nr(B_A, B_R)B_A + ne(B_A, B_R)B_R.$$

In this function, increasing a bonus in one component will affect both rates but will not affect the other bonus.[6] This function can be used to show the costs of different AC and RC bonus pairs, were they chosen.

We use our empirical results to compute AC and RC bonus cost estimates per 100 service members at an AC reenlistment point. We also compute the average cost per additional AC reenlistment and RC enlistment relative to a situation in which no bonuses are paid. The costs are shown in Tables 5.1–5.4. The tables show the increase in AC reenlistment (RC enlistment) as the AC (RC) bonus increases. The tables also show the average cost per additional reenlistment or enlistment.

The lower portions of the upper panels show our estimates of costs per additional AC reenlistment. These are in the range of $50,000–$65,000 for the Army, $25,000–$40,000 for the Navy, $17,000–$30,000 for the Marine Corps, and $130,000–$180,000 for the Air Force. The high estimates for Air Force costs reflect its low estimated responsiveness to AC bonuses. The lower panels show estimates of costs per additional RC enlistment, and these are in the range of $40,000–$65,000 for the Army, $30,000–$45,000 for the Navy, $25,000–$35,000 for the Marine Corps, and $30,000–$45,000 for the Air Force.

In these tables, as one bonus is changed the other bonus is held constant, so the change in bonus cost comes only from the bonus that changed. But as mentioned, the change in one bonus causes a change in both rates. A bonus increase results in an increased own rate and a decreased other rate. For instance, our estimates in Table 5.1 (Army) indicate that for a RC bonus of $4,000, an increase in the AC bonus from $4,000 to $8,000 increases AC reenlistments from 37.36 to 40.33, increases the AC bonus cost from $149,000 to $323,000, and increases the cost per additional AC reenlistment (relative to the base point of no AC or RC bonuses) from $49,000 to $53,000. The problem with these estimates is that they say nothing about what happens to the RC enlistment rate when the AC bonus increases. Looking back at our estimates in Table 4.2, we see that for an RC bonus of $4,000, an increase in the AC bonus from $4,000 to $8,000 decreases the RC enlistment rate from 0.257 to 0.245. If the RC wants to hold its rate at 0.257, it must increase its bonus. The second perspective, discussed next, shows the AC and RC bonus amounts and costs associated with increasing either rate while holding the other rate constant.

---

[6] However, if this function were in the joint bonus-setting model above, this statement would no longer be true. In that model, suppose that some external factor changed, for example, resulting in a shift in the AC marginal value curve or a shift in the AC supply function, i.e., the reenlistment probability curve. This would change the equilibrium values of both the AC and RC bonuses.

Table 5.1
Army AC and RC Bonus Cost

AC Bonus	RC Bonus				
	$0	$2,000	$4,000	$6,000	$8,000
Reenlistments per 100 service members at AC reenlistment point					
$0	35.35	34.76	34.29	33.96	33.76
$4,000	38.46	37.84	37.36	37.01	36.80
$8,000	41.46	40.83	40.33	39.97	39.76
$12,000	44.32	43.68	43.17	42.80	42.59
$16,000	47.01	46.37	45.85	45.48	45.27
Additional reenlistments relative to zero AC and RC bonuses					
$0	0	−0.59	−1.06	−1.39	−1.59
$4,000	3.10	2.49	2.01	1.66	1.45
$8,000	6.10	5.48	4.98	4.62	4.41
$12,000	8.97	8.33	7.82	7.45	7.24
$16,000	11.66	11.02	10.50	10.13	9.91
AC bonus cost ($ thousands)					
$0	0	0	0	0	0
$4,000	154	151	149	148	147
$8,000	332	327	323	320	318
$12,000	532	524	518	514	511
$16,000	752	742	734	728	724
Bonus cost per additional AC reenlistment ($ thousands)					
$4,000	50	49	49	49	48
$8,000	54	54	53	53	53
$12,000	59	59	58	58	58
$16,000	64	64	63	63	63

Table 5.1—Continued

RC Bonus	AC Bonus				
	$0	$4,000	$8,000	$12,000	$16,000
RC enlistments per 100 service members at AC reenlistment point					
$0	24.72	23.54	22.39	21.29	20.26
$2,000	25.98	24.75	23.56	22.43	21.36
$4,000	26.98	25.72	24.50	23.33	22.23
$6,000	27.69	26.41	25.17	23.98	22.86
$8,000	28.11	26.82	25.57	24.36	23.23
Additional enlistments relative to zero AC and RC bonuses					
$0	0	−1.19	−2.33	−3.43	−4.46
$2,000	1.26	0.03	−1.16	−2.29	−3.37
$4,000	2.25	1.00	−0.22	−1.39	−2.49
$6,000	2.97	1.69	0.45	−0.74	−1.86
$8,000	3.39	2.10	0.84	−0.36	−1.49
RC bonus cost ($ thousands)					
$0	0	0	0	0	0
$2,000	52	50	47	45	43
$4,000	108	103	98	93	89
$6,000	166	158	151	144	137
$8,000	225	215	205	195	186
Bonus cost per additional RC enlistment ($ thousands)					
$2,000	41	41	40	39	39
$4,000	48	47	46	46	45
$6,000	56	55	54	54	53
$8,000	66	65	65	64	63

NOTE: The bonus costs per additional RC enlistment are probably biased upward at higher levels of the RC bonus, as discussed in Chapter Four.

Table 5.2
Navy AC and RC Bonus Cost

AC Bonus	RC Bonus				
	$0	$2,000	$4,000	$6,000	$8,000
Reenlistments per 100 service members at AC reenlistment point					
$0	28.23	27.79	27.42	27.10	26.86
$4,000	33.25	32.77	32.36	32.01	31.74
$8,000	37.76	37.26	36.82	36.45	36.16
$12,000	41.56	41.04	40.58	40.20	39.90
$16,000	44.50	43.97	43.50	43.11	42.81
Additional reenlistments relative to zero AC and RC bonuses					
$0	0	−0.43	−0.81	−1.13	−1.37
$4,000	5.02	4.54	4.13	3.78	3.51
$8,000	9.54	9.03	8.59	8.22	7.93
$12,000	13.33	12.81	12.35	11.97	11.67
$16,000	16.27	15.74	15.27	14.88	14.58
AC bonus cost ($ thousands)					
$0	0	0	0	0	0
$4,000	133	131	129	128	127
$8,000	302	298	295	292	289
$12,000	499	492	487	482	479
$16,000	712	703	696	690	685
Bonus cost per additional AC reenlistment ($ thousands)					
$4,000	26	26	26	26	26
$8,000	32	32	31	31	31
$12,000	37	37	37	37	37
$16,000	44	43	43	43	43

Table 5.2—Continued

RC Bonus	AC Bonus				
	$0	$4,000	$8,000	$12,000	$16,000
RC enlistments per 100 service members at AC reenlistment point					
$0	17.70	16.47	15.35	14.42	13.69
$2,000	18.97	17.66	16.48	15.49	14.72
$4,000	20.08	18.71	17.47	16.43	15.63
$6,000	20.99	19.58	18.30	17.22	16.38
$8,000	21.70	20.25	18.94	17.83	16.97
Additional enlistments relative to zero AC and RC bonuses					
$0	0	−1.24	−2.35	−3.29	−4.01
$2,000	1.27	−0.04	−1.22	−2.21	−2.98
$4,000	2.37	1.00	−0.23	−1.27	−2.08
$6,000	3.29	1.88	0.60	−0.48	−1.32
$8,000	4.00	2.55	1.24	0.13	−0.73
RC bonus cost ($ thousands)					
$0	0	0	0	0	0
$2,000	38	35	33	31	29
$4,000	80	75	70	66	63
$6,000	126	117	110	103	98
$8,000	174	162	152	143	136
Bonus cost per additional RC enlistment ($ thousands)					
$2,000	30	29	29	29	28
$4,000	34	33	33	33	33
$6,000	38	38	37	37	36
$8,000	44	43	42	42	41

NOTE: The bonus costs per additional RC enlistment are probably biased upward at higher levels of the RC bonus, as discussed in Chapter Four.

Table 5.3
Marine Corps AC and RC Bonus Cost

AC Bonus	RC Bonus				
	$0	$2,000	$4,000	$6,000	$8,000
Reenlistments per 100 service members at AC reenlistment point					
$0	18.85	18.67	18.51	18.36	18.22
$4,000	24.46	24.25	24.05	23.86	23.70
$8,000	29.92	29.68	29.45	29.25	29.06
$12,000	34.75	34.50	34.25	34.02	33.82
$16,000	38.60	38.33	38.08	37.84	37.63
Additional reenlistments relative to zero AC and RC bonuses					
$0	0	−0.17	−0.34	−0.49	−0.63
$4,000	5.61	5.40	5.20	5.02	4.85
$8,000	11.07	10.83	10.61	10.40	10.21
$12,000	15.90	15.65	15.40	15.18	14.97
$16,000	19.75	19.48	19.23	18.99	18.78
AC bonus cost ($ thousands)					
$0	0	0	0	0	0
$4,000	98	97	96	95	95
$8,000	239	237	236	234	232
$12,000	417	414	411	408	406
$16,000	618	613	609	605	602
Bonus cost per additional AC reenlistment ($ thousands)					
$4,000	17	17	17	17	17
$8,000	22	22	22	21	21
$12,000	26	26	26	26	26
$16,000	31	31	31	31	31

Table 5.3—Continued

RC Bonus	AC Bonus				
	$0	**$4,000**	**$8,000**	**$12,000**	**$16,000**
RC enlistments per 100 service members at AC reenlistment point					
$0	10.51	9.78	9.07	8.45	7.95
$2,000	11.33	10.56	9.80	9.13	8.59
$4,000	12.11	11.29	10.49	9.77	9.20
$6,000	12.83	11.97	11.12	10.37	9.77
$8,000	13.48	12.57	11.69	10.90	10.28
Additional enlistments relative to zero AC and RC bonuses					
$0	0	−0.73	−1.43	−2.06	−2.56
$2,000	0.83	0.05	−0.71	−1.38	−1.91
$4,000	1.61	0.78	−0.02	−0.73	−1.30
$6,000	2.32	1.46	0.61	−0.14	−0.74
$8,000	2.97	2.07	1.18	0.40	−0.23
RC bonus cost ($ thousands)					
$0	0	0	0	0	0
$2,000	23	21	20	18	17
$4,000	48	45	42	39	37
$6,000	77	72	67	62	59
$8,000	108	101	94	87	82
Bonus cost per additional RC enlistment ($ thousands)					
$2,000	28	27	28	26	26
$4,000	30	30	30	29	29
$6,000	33	33	33	32	32
$8,000	36	36	36	35	35

NOTES: The bonus costs per additional AC reenlistment in the upper panel are estimated correctly. They are all the same when rounded to the nearest dollar. The bonus costs per additional RC enlistment are probably biased upward at higher levels of the RC bonus, as discussed in Chapter Four.

Table 5.4
Air Force AC and RC Bonus Cost

AC Bonus	RC Bonus				
	0	$2,000	$4,000	$6,000	$8,000
Reenlistments per 100 service members at AC reenlistment point					
$0	48.96	48.27	47.68	47.19	46.82
$4,000	50.43	49.74	49.15	48.65	48.28
$8,000	51.72	51.04	50.44	49.95	49.58
$12,000	52.85	52.17	51.57	51.08	50.71
$16,000	53.80	53.12	52.53	52.04	51.66
Additional reenlistments relative to zero AC and RC bonuses					
$0	0	−0.68	−1.28	−1.77	−2.14
$4,000	1.47	0.78	0.19	−0.30	−0.68
$8,000	2.76	2.08	1.49	0.99	0.62
$12,000	3.89	3.21	2.61	2.12	1.75
$16,000	4.85	4.16	3.57	3.08	2.71
AC bonus cost ($ thousands)					
$0	0	0	0	0	0
$4,000	202	199	197	195	193
$8,000	414	408	404	400	397
$12,000	634	626	619	613	608
$16,000	861	850	840	833	827
Bonus cost per additional AC reenlistment ($ thousands)					
$4,000	137	136	134	133	132
$8,000	150	148	146	145	144
$12,000	163	161	159	158	156
$16,000	178	176	173	172	171

Table 5.4—Continued

RC Bonus	AC Bonus				
	0	$4,000	$8,000	$12,000	$16,000
RC enlistments per 100 service members at AC reenlistment point					
$0	16.81	16.32	15.90	15.52	15.21
$2,000	17.97	17.46	17.01	16.62	16.28
$4,000	18.98	18.45	17.98	17.57	17.22
$6,000	19.81	19.26	18.77	18.35	17.99
$8,000	20.45	19.88	19.38	18.95	18.58
Additional enlistments relative to zero AC and RC bonuses					
$0	0	−0.48	−0.91	−1.28	−1.60
$2,000	1.16	0.65	0.20	−0.19	−0.52
$4,000	2.17	1.64	1.17	0.76	0.41
$6,000	3.01	2.46	1.97	1.55	1.19
$8,000	3.64	3.08	2.58	2.15	1.78
RC bonus cost ($ thousands)					
$0	0	0	0	0	0
$2,000	36	35	34	33	33
$4,000	76	74	72	70	69
$6,000	119	116	113	110	108
$8,000	164	159	155	152	149
Bonus cost per additional RC enlistment ($ thousands)					
$2,000	31	31	31	30	31
$4,000	35	35	35	34	34
$6,000	40	39	39	39	39
$8,000	45	45	44	44	44

Note: The bonus costs per additional RC enlistment are probably biased upward at higher levels of the RC bonus, as discussed in Chapter Four.

Bonus Cost with Respect to AC Reenlistment and RC Enlistment Rates

The joint cost function can be used to identify the own incremental bonus cost and the bonus cost imposed on the other component to hold its rate constant. The joint cost function can be defined in terms of AC and RC bonuses as just discussed or in terms of AC and RC rates. The joint cost function in terms of rates is

$$nrB_A(r,e) + neB_R(r,e).$$

The bonus functions $B_A(r,e)$ and $B_R(r,e)$ are obtained by inverting the probability functions, i.e., by solving for the bonuses as functions of the probabilities. The bonus functions for our empirical model are given above in the discussion of bonus setting with respect to target rates.[7] We have used these functions along with our empirical estimates of bonus effects to compute AC and RC bonuses and bonus costs for a range of AC reenlistment rates and RC enlistment rates. The results are shown in Tables 5.5–5.8.[8]

The tables show, as expected, that the own bonus increases as the own rate increases. For instance, Table 5.5 for the Army indicates that an increase in the reenlistment rate from 0.39 to 0.40 requires an increase in the reenlistment bonus from $5,281 to $6,970 when the RC enlistment rate is 0.24. Further, our estimates suggest that to hold the RC rate at 0.24, the RC bonus must increase from $1,347 to $2,256. The AC bonus increase causes AC bonus cost to increase from $206,000 to $279,000 and RC bonus cost to increase from $32,000 to $54,000. Further, for every 100 soldiers at an AC reenlistment point, increasing the reenlistment rate from 0.39 to 0.40 produces one additional reenlistment, and the marginal cost of this is $279,000 – $206,000 = $73,000 for the AC and $54,000 – $32,000 = $22,000 for the RC. Depending on bonus budgeting procedures, the AC might bear a cost of $73,000 from its bonus budget, and the RC might either bear the cost of $22,000 from its budget or

---

[7] Although this approach is general, the quadratic functional form we have chosen for our probabilities is not ideal. The bonus functions include a square-root term, and at certain values of the reenlistment and enlistment rates the quantities in this term can be negative. This is problematic because the square root of a negative number is an imaginary number. This condition occurs when the implied value of the bonus is greater than the point at which the bonus has its maximum effect (see Figures 4.1–4.4), i.e., when further increases in the bonus are associated with a lower, not a higher, rate. Therefore, given our empirical specification, meaningful values of the bonuses and bonus costs are found only in the range below the maximum bonus effect.

[8] Tables 5.5–5.8 show the bonuses and bonus costs associated with different AC reenlistment and RC enlistment rates. They should not be interpreted as equilibrium rates; although any of the rates could feasibly be equilibrium rates, without knowing the AC and RC value functions we cannot say where the equilibrium would occur. The table should be interpreted as saying that *if* the equilibrium rates were some pair (*r*,*e*), the bonuses shown in the table for that pair would be the equilibrium bonuses.

Table 5.5
Army AC and RC Bonuses and Bonus Costs, by Rates

AC Reenlistment Rate	RC Enlistment Rate				
	.230	.235	.240	.245	.250
AC bonus ($)					
.38	2,845	3,238	3,639	4,048	4,465
.39	4,444	4,858	5,281	5,712	6,153
.40	6,086	6,523	6,970	7,426	7,892
.41	7,778	8,240	8,713	9,196	9,691
.42	9,525	10,015	10,517	11,031	11,557
RC bonus ($)					
.38	0	0	549	1,576	2,744
.39	0	328	1,347	2,500	3,864
.40	104	1,117	2,256	3,592	5,301
.41	884	2,011	3,322	4,969	7,628
.42	1,766	3,055	4,648	7,040	> 7,040
AC bonus cost ($ thousands)					
.38	108	123	138	154	170
.39	173	189	206	223	240
.40	243	261	279	297	316
.41	319	338	357	377	397
.42	400	421	442	463	485
RC bonus cost ($ thousands)					
.38	0	0	13	39	69
.39	0	8	32	61	97
.40	2	26	54	88	133
.41	20	47	80	122	191
.42	41	72	112	172	> 172

NOTE: The RC bonus costs are probably biased upward at higher levels of the RC enlistment rate, as discussed in Chapter Four.

Table 5.6
Navy AC and RC Bonuses and Bonus Costs, by Rates

AC Reenlistment Rate	RC Enlistment Rate				
	.150	.155	.160	.165	.170
AC bonus ($)					
.38	8,084	8,327	8,576	8,831	9,093
.39	9,189	9,454	9,725	10,005	10,292
.40	10,366	10,656	10,956	11,265	11,584
.41	11,632	11,956	12,291	12,638	12,998
.42	13,016	13,382	13,764	14,163	14,580
RC bonus ($)					
.38	0	396	1,394	2,464	3,631
.39	0	896	1,945	3,081	4,339
.40	396	1,429	2,538	3,755	5,129
.41	913	2,001	3,182	4,499	6,029
.42	1,466	2,618	3,888	5,336	7,092
AC bonus cost ($ thousands)					
.38	307	316	326	336	346
.39	358	369	379	390	401
.40	415	426	438	451	463
.41	477	490	504	518	533
.42	547	562	578	595	612
RC bonus cost ($ thousands)					
.38	0	6	22	41	62
.39	0	14	31	51	74
.40	6	22	41	62	87
.41	14	31	51	74	103
.42	22	41	62	88	121

NOTE: The RC bonus costs are probably biased upward at higher levels of the RC enlistment rate, as discussed in Chapter Four.

Table 5.7
Marine Corps AC and RC Bonuses and Bonus Costs, by Rates

AC Reenlistment Rate	RC Enlistment Rate				
	.100	.105	.110	.115	.120
AC bonus ($)					
.29	7,507	7,637	7,768	7,901	8,036
.30	8,314	8,451	8,591	8,733	8,877
.31	9,144	9,291	9,440	9,592	9,747
.32	10,004	10,162	10,322	10,485	10,652
.33	10,900	11,070	11,243	11,420	11,601
RC bonus ($)					
.29	2,314	3,833	5,456	7,232	9,253
.30	2,739	4,308	5,994	7,862	10,033
.31	3,184	4,806	6,566	8,543	10,906
.32	3,650	5,332	7,177	9,285	11,905
.33	4,139	5,889	7,832	10,105	13,098
AC bonus cost ($ thousands)					
.29	218	221	225	229	233
.30	249	254	258	262	266
.31	283	288	293	297	302
.32	320	325	330	336	341
.33	360	365	371	377	383
RC bonus cost ($ thousands)					
.29	23	40	60	83	111
.30	27	45	66	90	120
.31	32	50	72	98	131
.32	36	56	79	107	143
.33	41	62	86	116	157

NOTE: The RC bonus costs are probably biased upward at higher levels of the RC enlistment rate, as discussed in Chapter Four.

Table 5.8
Air Force AC and RC Bonuses and Bonus Costs, by Rates

AC Reenlistment Rate	RC Enlistment Rate				
	.230	.235	.240	.245	.250
AC bonus ($)					
.38	5,398	6,573	7,816	9,137	10,546
.39	7,349	8,627	9,987	11,444	13,018
.40	9,453	10,857	12,367	14,004	15,800
.41	11,749	13,316	15,026	16,914	19,042
.42	14,297	16,087	18,084	20,365	23,073
RC bonus ($)					
.38	0	792	1,949	3,228	4,688
.39	67	1,174	2,381	3,729	5,298
.40	427	1,576	2,840	4,270	5,974
.41	803	2,000	3,328	4,857	6,739
.42	1,199	2,449	3,853	5,502	7,635
AC bonus cost ($ thousands)					
.38	275	335	399	466	538
.39	378	444	514	589	670
.40	492	565	643	728	822
.41	617	699	789	888	1,000
.42	758	853	958	1,079	1,223
RC bonus cost ($ thousands)					
.38	0	13	33	56	84
.39	1	19	40	65	95
.40	7	26	48	75	108
.41	13	33	57	85	121
.42	19	40	65	96	137

NOTE: The RC bonus costs are probably biased upward at higher levels of the RC enlistment rate, as discussed in Chapter Four.

suffer a decrease in RC enlistment. The potential decrease in enlistment can be found by using the supply equations.[9]

We can also compute an average cost per additional reenlistment as we did in Tables 4.2–4.6. In this case, we take the low value in our reenlistment rate range as the base point. This value is 0.38. The move from 0.38 to 0.40 results in two additional reenlistments per 100 soldiers at the reenlistment point. The AC bonus cost increases from $138,000 to $279,000, or a cost per additional reenlistment of ($279,000 – $138,000)/2 = $70,500. The RC bonus cost increases from $13,000 to $54,000. Including the RC cost increases the cost per additional reenlistment to $91,000.

The computations are similar for an increase in the RC enlistment rate while holding the AC reenlistment rate constant. As an example, consider a one-point increase in the Navy RC enlistment rate from 0.155 to 0.165 at an AC reenlistment rate of 0.40 (Table 5.6). Our estimates indicate that to accomplish this, the RC bonus increases from $1,429 to $3,755 and the AC bonus increases from $10,656 to $11,265. The RC bonus cost increases from $22,000 to $62,000, or by $40,000, and the AC bonus cost increases from $426,000 to $451,000, or by $25,000.

Summary

We have discussed two models of bonus setting: (1) setting bonuses to reach target reenlistment and enlistment rates and (2) setting bonuses in a notional framework to maximize the value of additional reenlistments or enlistments relative to bonus costs. We have presented two models of bonus cost, one in terms of AC and RC bonuses and the other in terms of AC and RC reenlistment and enlistment rates. We have related the concepts in the bonus-setting models to the bonus cost models, and we have used the bonus cost models along with our empirical estimates to reveal how AC and RC rates and bonus costs change as bonuses change, and how AC and RC bonuses and bonus costs change as rates change. In addition to quantifying these relationships, we have shown how changing one bonus but not the other can reduce the other rate, and how changing one rate while holding the other rate constant requires an increase in both bonuses and hence in both bonuses' costs. The tables in this chapter give estimates of the costs from both perspectives.

---

[9]   From Table 5.5, we see that the initial rates of $r = 0.39$ and $e = 0.24$ are supported by bonuses of $5,281 and $1,347, respectively. The supply equations show that if the AC bonus increases to $6,970 but the RC bonus is unchanged, the RC enlistment rate decreases to 0.235. Further, because the RC bonus has not changed and the RC rate has been allowed to decrease, the AC reenlistment rate actually becomes 0.403, i.e., slightly higher than 0.40.

CHAPTER SIX

Conclusion

We are able to draw a number of conclusions from our findings. First, reserve enlistment and affiliation bonuses have a positive impact on prior service accession rates in the Selected Reserve. Policy changes taking effect in 2006 made these reserve bonuses a more flexible tool by increasing eligibility and bonus ceilings, and the subsequent expansion in reserve bonus use and levels helped the reserves meet their accession goals at a time when the Army Guard and Reserve and the Marine Corps Reserve faced shortages.

Second, we find that AC members at the point of reenlistment consider and respond to both active and reserve bonuses, and this supply-side behavior implies that each component imposes an externality on the other when setting bonuses. Higher AC bonuses increase AC reenlistment and decrease RC enlistment, and higher RC bonuses increase RC enlistment and decrease AC reenlistment. That is, there is a positive own-effect and a negative cross-effect. It can be shown analytically that increasing either bonus keeps more people in the military overall, i.e., in the AC plus RC.[1] But the negative cross-effect means that a higher AC bonus increases AC reenlistment partly at the expense of lower RC enlistment, and similarly for the effect of a higher RC bonus. As we have shown, if an RC bonus is held constant while an AC bonus is increased, RC enlistment decreases. In Chapter Five, we computed the amount by which the RC bonus must be increased to hold the RC rate constant. This calculation allows us to quantify the value of the externality imposed on the RC when the AC raises its bonus. For example, if the Army wants to increase its AC reenlistment rate from 0.39 to 0.41 and hold its RC enlistment rate at 0.24, RC bonus expenditures must increase from $32,000 to $80,000, or by nearly $50,000, per 100 AC members at reenlistment. Further, if the externalities were not taken into account, the interaction between AC and RC bonus effects might lead to unexpected decreases in RC enlistment, or AC reenlistment, or to the appearance that AC and RC bonuses were ineffective.

---

[1] For instance, the total force effect of a higher AC bonus on AC reenlistment and RC enlistment equals $f'(b)r(1-r-e)$ where $f'(b)$ is the effect of the bonus on the probability of AC reenlistment, r is the AC reenlistment probability, and e is the RC enlistment probability.

Third, given the interaction between active and reserve bonuses, both components would benefit from coordinated bonus setting. The interaction of bonus effects is a fundamental part of the market; it is a relationship that stems from supply behavior and not an artifact that will disappear. Our study did not delve into recruiter behavior and whether the AC and RC actively compete for recruits, though the supply-side behavior we have estimated means that they compete implicitly if not explicitly. We also have not studied whether either component is aware of the impact of its bonus decisions on the other. But as a general point, awareness of the cross-effects and communication between AC and RC bonus setters can help to efficiently coordinate bonuses and ensure that bonus budgets are set appropriately and used efficiently.

Fourth, our estimates indicate that AC deployment has a small-to-negligible overall effect on RC enlistment. An exception is Army short (1–9 months) hostile deployment, which we find significantly increased AC reenlistment and decreased RC enlistment.[2] The higher bonuses we observe for the RC Army helped to offset the negative impact of short hostile deployment on RC enlistment. To explore this further, we ran models to test whether the estimated bonus effects were sensitive to the amount of deployment and found no support for this hypothesis. The effectiveness of the bonuses did not depend on the amount of the individual's AC deployment.

Finally, RC prior service enlistment rates have trended downward since 1999, even after controlling for all covariates in our model. Although the year effects are somewhat jagged, they suggest a decrease in the RC enlistment rate of about 7 percentage points from 2001 to 2008.[3] Possible reasons for the downward trend include an improving economy throughout most of the 2000s and a reaction to increased deployments of RC units. As mentioned in Chapter Four, the propensity of youth to enlist in the military also decreased over this period, and it has been suggested that the decrease was a response to the increasing likelihood of deployment (Asch et al., 2010).

In our model, the downward trend represents a shift in the reenlistment and enlistment supply curves. To illustrate, suppose the AC and RC Army bonuses were $9,000 and $1,400 and the reenlistment and enlistment rates were 0.40 and 0.25. Suppose each rate fell by a point, to 0.39 and 0.24, because of the downward trend.[4] To

---

[2]   Recall that the AC deployment variable enters both the AC reenlistment and RC enlistment probability expressions.

[3]   The domestic economy was strong and the unemployment rate improved during this period. Although our study does not isolate the effect of unemployment on AC reenlistment or RC enlistment, higher unemployment tends to increase AC recruiting and retention and may also increase RC recruiting and retention. Participation in the RC provides a stable source of income that may be all the more attractive as the risk of unemployment increases. Therefore, we suspect the decrease in unemployment from 2003 to 2007 contributed to the downward trend in RC enlistment. By the same token, the current severe recession should improve AC and RC recruiting and retention.

[4]   In our example, the downward shift in the curves is characterized by a decrease in the intercepts c_1 and c_2 discussed in Chapter Four.

restore the rates to their initial levels, the bonuses would have to increase to $11,800 and $5,400. The relatively large increase in the RC bonus is necessary because of the downtrend itself and the strong negative cross-effect on RC enlistment as the AC bonus increases.

More broadly, our findings shed light on historical RC enlistment and bonus trends. In particular, the large increase in RC bonuses starting in 2005 can be interpreted as a policy response to bonus increases by the AC. Chapter Two documents the increase in AC service members in the Army and Marine Corps that received a reenlistment bonus in 2005–2008 relative to 2003 and 2004, and the increase in the average amount of the bonus. We also noted the 2006 legislation that increased RC enlistment and affiliation bonus eligibility and bonus ceilings. These changes were preceded by an increase in the percentage of RC prior service enlistments that received a bonus and the average bonus amount that began in 2005, the year before the legislative changes, and continued through 2008. In the context of our model and findings, the RC bonus changes can be seen as a response to the growth in AC bonuses; given an increase in the average AC bonus, the RC bonus had to increase to prevent the RC enlistment rate from decreasing. For instance, consider the average AC Army first-term bonus, which in our sample increased from approximately $9,000 in 2004 to $12,000 in 2005 and $15,000 in 2006 (Figure 2.4). The average RC Army bonus increased from about $1,400 in 2004 to $6,000 in 2005 and $7,000 in 2006 (Figure 2.4). Our estimates suggest that at an RC bonus of $1,400, an increase in the AC bonus from $9,000 to $15,000 would increase the AC reenlistment rate from 0.42 to 0.46 and decrease the RC enlistment rate from 0.23 to 0.21. With the AC bonus at $15,000, an increase in the RC bonus from $1,400 to $7,000 would decrease the AC rate to 0.45 and increase the RC rate back to 0.23, its initial level.

Our findings also suggest that the expanded use and generosity of RC Army bonuses helped to restore and maintain the RC Army enlistment rate as the AC Army stabilized its reenlistment rate and pursued force growth. Hosek and Martorell (2009) provide evidence that the Army AC reenlistment bonus increases starting in 2005 were critical to sustaining its reenlistment rate in the face of downward pressure from extensive deployments. The AC Army was also ordered to grow (Table 2.2), and the greater use of bonuses helped it to grow by preventing a decrease in reenlistment. The surge in reenlistment in 2003 and 2004 following the defeat of Saddam's army, the continued activation and deployment of RC units, and the strong civilian economy probably served to decrease the personnel flow from the AC to the RC and contributed to the decrease in RC Army strength in 2004 and 2005. During these and subsequent years, the RC Army was not growing—Army National Guard and Army Reserve authorized strengths remained constant (Table 2.2). If the RC Army had been growing, shortages and pressure to respond would have been even greater. Our analysis suggests that the increase in RC Army bonuses from 2005 onward offset the negative effect on RC enlistment that would have occurred as a side effect of AC Army bonus usage as it pursued its growth mandate. Further, to the extent that expectations of RC deployment negatively affected RC enlistment, the greater use of RC bonuses was a helpful measure.

Looking to the future, we see that the emergence of the RC as strategic and operational force carries an implication for defense manpower research. For force planners, programmers, and commanders, there is a growing stake in knowing about the effects of policies intended to affect reserve recruiting, retention, compensation, motivation, morale, retirement, family support, training, and leadership. More work needs to be done to address these concerns. As a foundation for analysis, it will be important to extend and develop databases from defense administrative files and surveys and to link these data to other relevant data, such as Social Security earnings data. Our study is an example of research relevant to the new reserves, and we have found a previously unrecognized connection between AC and RC bonuses. In the background, we also found that available data on AC and RC recruiting resources are limited. There are no individual-level files of AC and RC bonus offerings; we had to impute bonuses, which added to issues in identifying bonus effects. Nor are there individual-level data on the objectives and incentives of AC career counselors to "sell" RC enlistment in addition to AC reenlistment. Finally, data on RC recruiting resources, recruiting goals, recruiter incentives, advertising, and educational benefits are limited in comparison to those for the AC. Investing in the development of RC data will strengthen the capability to address a range of defense manpower issues as they increasingly parallel those of the active components in substantive content and policy importance. Data on RC bonuses, counseling, and recruiting would, for example, help to support analysis aimed at how bonus effects differ across subgroups of the military population. Such information would be useful in targeting bonuses more efficiently.

Data

This appendix describes the data sets and the creation of the analysis file, including the construction of variables for AC reenlistment and RC enlistment, AC and RC bonus variables, AC deployment, and demographic variables. This appendix also includes tables of means for the analytical sample versus the full data for the first-term and the second-and-higher-term data, as well as tables that show estimates using three- and 12-month windows rather than the one-month window used for the estimates in main text.

Data Sets

We employ the following DMDC databases:

Proxy Perstempo File. DMDC's Proxy Perstempo file is our primary data source on reenlistment decisions, deployments, and service member characteristics. For all AC service members, the file contains individual-level records with longitudinal information on branch of service, time remaining on the current enlistment contract, MOS, pay grade, and deployment history. The file also contains demographic information, including family status, education, gender, date of birth, race, ethnicity, and AFQT scores.

Joint Uniform Military Pay System (JUMPS) and the Reserve Pay File (RPF). The JUMPS file is our primary data source for information on AC reenlistment bonuses. The file contains information on all payments made to active-duty personnel for each month, including both regular and bonus pays. The records indicate the amount of reenlistment bonus paid out in each month, as well as the total bonus amount associated with each bonus payout.

The RPF is our primary data source for information on RC enlistment and affiliation bonuses. The RPF is the RC analog to the JUMPS. Like JUMPS, the file contains information on all payments made to reserve personnel for each month, including both regular and bonus pays. However, the file does not include information about the total bonus amount associated with a given bonus payout.

Creation of Analysis Data File

Identifying Reenlistment Decisions. Because we do not have access to reenlistment contract records, we followed work by Hosek and Totten (2002) and Hosek and Martorell (2009) by identifying reenlistment decision points based on information on expiration of term of service (ETS) associated with the current contract. In particular, using the Proxy Perstempo file, we identified reenlistment points as instances where ETS increased by 24 or more months in a single month, thus indicating that the service member had signed a new contract, or when ETS went to zero, thus indicating that the service member had left the AC.

We coded those active-duty service members whose ETS increased by 24 or more months as reenlisting in the AC. To identify RC enlistments, we tracked former active-duty service members (i.e., those whose ETS went to zero) into the RPF, and we coded those receiving pay from the Selected Reserves within the two-year period following their exit from active duty as having joined the RC. We coded the remaining service members who had left the AC and failed to join the RC within two years as having left the military. Roughly 90 percent of all AC personnel who join the RC after their first term do so within two years, thus suggesting that our follow-up window is sufficient (Hosek and Martorell, 2009).

Having defined reenlistment decisions in this way, we searched the Proxy Perstempo file for all instances of reenlistment decision points from 1999 through June 2009 and tracked those individuals who exited the AC into the RPF to identify whether or not they subsequently joined the RC.

We make two additional sample restrictions. First, we dropped individuals with missing demographic information. Second, we excluded service members with less than three years of AC service at the time of their first reenlistment decision point.[1] We dropped these individuals because our analysis uses information about deployments during the three years prior to the reenlistment decision, and we did not have that information for early reenlisters.

Deployment Measures. Our primary data source to identify deployment histories is the Proxy Perstempo file. The file has monthly information on current deployment status and receipt of hostile fire pay. We use this information to cumulate total months of deployment and total months with hostile fire pay in the 36-month period ending three months prior to the reenlistment decision point. The three-month buffer guards against possible complications stemming from service members choosing to reenlist while on deployment in order to take advantage of the combat zone tax exclusion.

Bonus Data. Our data on AC reenlistment bonuses come from JUMPS, and our data on RC enlistment and affiliation bonuses come from the RPF. Our economet-

---

[1]   There are 5,606,296 reenlistment decision points for service members with demographic information in our database. Eliminating service members with less than three years of AC service eliminates 1,125,563 decisions, leaving us with 4,480,663 individual reenlistment decisions.

ric model requires information about the bonus offered by each component. Neither data set we use for our bonus data has information on bonus offers—we are only able to observe bonus pays. Given that the AC and RC bonus variables are constructed differently, as discussed below, we allow for their effects to differ in our empirical specification.

We follow Hosek and Martorell (2009) by imputing bonus offers based on information on bonus pays for similar service members at the time of reenlistment. For each cell determined by service, zone, pay grade, occupation (using each service's occupation coding), month of reenlistment decision, and year of decision, we calculate the mean bonus paid in each component conditional upon joining that component and receiving a bonus.[2] We coded personnel falling in cells in which fewer than 15 percent of service members who joined a component received a bonus as receiving a zero bonus. Otherwise, we coded them as receiving a bonus offer equal to the mean within their cell.

There are several points to note about our bonus imputations. First, our method imputes an expected bonus within a cell. For AC reenlistments, most stay within the same occupational specialty and face the same bonus offer. However, RC enlistments may be into the same or a different occupational specialty, not all of which offer the same bonus. The expected value of the RC bonus is therefore a mix of bonus offers across the RC specialties chosen by members within an AC specialty who leave the AC, join the RC, and receive a bonus. Given that service members moving from the AC to the RC may change specialty, this method provides a rational expectation of their RC bonus offer. Therefore, in interpreting our results, we must keep in mind the relatively higher likelihood of occupational switching among those who join the RC relative to those who reenlist in the AC. Also, service members may not reenlist/join the RC immediately after their contract has expired, and if they wait long enough, the bonus may change.

Second, there are differences in the degree of bonus information in the JUMPS and RPF that necessitated slightly different calculation of bonus variables across components. The JUMPS provides information about the total contracted bonus amount associated with each bonus payout, but the RPF provides information only on month-by-month bonus payouts. Thus, for the AC, we were able to obtain the total bonus amount directly from the data. For the RC, we used OSD regulations to impute the total bonus. In particular, for all but service members with 8–10 years of prior service, both reenlistment and affiliation bonus payouts in the RC are made half up front and half in installments (Reserve Forces Almanacs). Thus, we imputed the total RC bonus as twice the initial bonus payout. As a specification check, we ran a number of our models for second-term personnel dropping service members with 8–10 years of prior service, and the results are qualitatively similar to those we report in the this docu-

---

[2]   We also ran models in which bonuses were coded as the unconditional mean within a cell, including zero values. The estimates are qualitatively similar.

ment. Nevertheless, this imputation could introduce more measurement error in RC bonuses than in AC bonuses and bias our estimates toward finding AC bonus impacts that are relatively larger than RC bonus impacts.

Third, our method requires us to restrict attention to cells in which a sufficient number of service members both reenlisted and joined the RC. Specifically, we restrict attention to cells in which at least five service members reenlisted and joined the RC. This restriction significantly reduces our sample size and affects the data means (see below). Restricting attention to cells with sufficiently large numbers of service members who both reenlist and join the RC also leads to two distinct forms of sample selection, as described below.

We are restricting attention to a priori large cells; that is, cells with large numbers of service members. This form of selection is largely innocuous, as we are selecting on an independent variable. As such, if this were the only form of selection, we would get unbiased estimates for service members in MOSs with large numbers of service members.

A more problematic form of sample selection occurs because, conditional on cell size, we are selecting on reenlistment and RC join rates. This form of selection on the dependent variable may bias our estimates because it effectively causes us to select on the error of the regression equation. This is because at any bonus value, observations with sufficiently large negative errors are eliminated from the sample. Thus, as reenlistment and RC join rates increase, the expected value of the error becomes increasingly positive, making these two elements positively correlated. Because this contradicts the standard assumption of ordinary least squares (OLS) that the error and the independent variables are not correlated, OLS estimates become biased. While this form of bias can be either positive or negative, it generally leads to attenuation of estimated effects (Hausman and Wise, 1977).

To address concerns over the latter form of selection bias, we also ran models where we selected solely on the number of AC service members in a cell. That is, we restricted attention to cells with a minimum number of AC service members, where that minimum was chosen so that all cells had a sufficient number of service members choosing both AC and RC.[3] In practice, this approach was complicated by the fact that many cells had either unusually low AC reenlistment or RC join rates, requiring us to set the minimum cell size at a high level. This reduces our sample size and increases the average cell size, thereby significantly reducing the total number of cells in the analysis. Since our key explanatory variables, the imputed bonus values, vary only at the cell level, this drastically reduces our effective sample size.

In practice, we found that using a one-month window was not possible with this approach, since the number of included cells was so low. We thus experimented with three- and 12-month cell windows. Moving to larger windows increases the average

---

[3]   We thank Paul Hogan for this suggestion.

cell size and increases the number of service members we are able to include in our analysis. However, since bonuses change on a monthly basis, using a larger window also increases the measurement error in our imputed bonuses, which can attenuate our estimates.

We present estimates that select on cell size using both three- and 12-month windows in Tables A.1 and A.2. While the estimates differ in magnitude from those that select on the number of service members reenlisting and joining the RC that we report in Chapter Four, they are qualitatively similar. Moreover, as expected, the results using 12-month windows are often attenuated relative to those using a three-month window.

Finally, there are several reasons why our imputed bonuses might not be equal to true bonus offers. First, while the bonus multiplier should be constant within a cell, actual bonus amounts will vary according to the term of reenlistment, with larger reenlistment periods garnering larger bonuses. Second, we define bonuses according to the occupation that service members had on active duty while they were nearing the end of their term of service. However, bonus multipliers are set according to service member occupation at the start of the new term of service. We should thus observe variance in bonus pays within a cell when service members switch occupations at the time of signing a new contract. We expect this to occur more when service members join the RC than when they reenlist in the AC, so we expect to see more within-cell variance in RC bonuses than in AC bonus.

To examine the degree of variance in bonus pays accounted for by our bonus imputations, we regressed observed bonus pays in the AC and RC on a full set of cell dummies (service, zone, pay grade, occupation, month of reenlistment decision, year of decision). For both the AC and the RC, this model yielded an adjusted R-square of approximately 0.77, indicating that 77 percent of the variance in bonus pays is accounted for by our bonus imputations.[4] The remaining 23 percent of variance in bonus pays occurs within cells.

To the extent that there is meaningful variance in bonus offers within cells, another concern is that actual observed bonuses may be a nonrepresentative sample of the complete set of offered bonuses. Since bonuses increase the probability of reenlistment, we expect those who reenlist to have higher bonus offers than those who do not. This would lead to a systematic upward bias in our bonus imputations, the extent of which is positively related to the impact of bonuses on reenlistment decisions. This form of bias in the bonus imputations leads to attenuation in our estimates of the effect of bonuses on reenlistment decisions. However, there are a number of reasons why we

---

[4] Given that most of the variance in bonus pays occurs between cells, we attempted to cluster our standard errors at the cell level. Unfortunately, the algorithm that STATA uses to cluster standard errors failed to converge, so we report standard errors that do not account for clustering at the cell level. These standard errors are likely biased downward. Nevertheless, clustering does not change our point estimates, and, given our sample sizes, our point estimates are so precise that accounting for clustering should not alter our conclusions regarding statistical significance.

Table A.1
Point Estimates Selecting on Cell Size: Three-Month Window

	Army	Navy	Marine Corps	Air Force
First term				
AC bonus	0.0344***	0.0639***	0.0899***	0.0155***
AC bonus squared	−0.0003*	−0.0012***	−0.0017***	−0.0002**
RC bonus	0.037***	0.0458***	0.0448	0.0439***
RC bonus squared	−0.0019***	−0.0018***	−0.0012	−0.0017***
	SE	SE	SE	SE
AC bonus	0.003	0.0024	0.0037	0.0024
AC bonus squared	0.0001	0.0001	0.0001	0.0001
RC bonus	0.0033	0.0082	0.0332	0.0084
RC bonus squared	0.0002	0.0004	0.0023	0.0005
Second+ term				
AC bonus	0.054***	0.0463***	0.0534***	0.0322***
AC bonus squared	−0.0009***	−0.0007***	−0.0009*	−0.0009***
RC bonus	0.0178***	0.0656***	−0.0413	0.0193
RC bonus squared	−0.0009**	−0.0031***	−0.0001	−0.0015*
	SE	SE	SE	SE
AC bonus	0.0028	0.0035	0.0098	0.0034
AC bonus squared	0.0001	0.0001	0.0004	0.0001
RC bonus	0.005	0.0127	0.0955	0.0132
RC bonus squared	0.0003	0.0007	0.0059	0.0008

NOTE: SE is the standard error of the coefficient. Statistical significance: *** = 0.001, ** = 0.01, * = 0.1.

are not particularly concerned with this form of bias. First, as we have shown, this bias should attenuate our results, so that we may still interpret our results as a lower-bound estimate of the effect of bonuses. Second, as noted above, most of the variance in bonuses occurs across, rather than within, cells. Finally, our bonus effect estimates are relatively modest, on the order of three percentage points for an $8,000 bonus, which represents nearly the full range of bonus offers. Since the reenlistment rates of those with large bonus offers are not much different than those with small bonus offers, there extent of this bias is likely to be small in practice.

Covariates. Our control variables come from the PERSTEMPO file. We link data on years of service, race, ethnicity, gender, marital status, MOS, AFQT score, and

Table A.2
Point Estimates Selecting on Cell Size: 12-Month Window

	Army	Navy	Marine Corps	Air Force
First term				
AC bonus	0.0343***	0.0327879***	0.1354084***	−0.0138335***
AC bonus squared	−0.0000261	−0.0003737***	−0.0027538***	0.0002767***
RC bonus	0.111977***	0.0617585***	0.1045579***	0.0480834***
RC bonus squared	−0.0052833***	−0.002642***	−0.0049091***	−0.0014016***
	SE	SE		SE
AC bonus	0.0038224	0.0022832	0.0025562	0.0016937
AC bonus squared	0.0001597	0.0000607	0.00008	0.0000488
RC bonus	0.0045092	0.005006	0.0155507	0.0056149
RC bonus squared	0.0002243	0.0002674	0.0010123	0.0003324
Second+ term				
AC bonus	0.0026132	0.0193068***	0.0385793***	0.0149101***
AC bonus squared	0.0009715***	−0.0002438***	−0.0008463***	−0.0005129***
RC bonus	0.0584824***	0.0463336***	0.0255389	0.0446072***
RC bonus squared	−0.0025953***	−0.0018658***	−0.002165	−0.002105***
	SE	SE	SE	SE
AC bonus	0.0025113	0.0026215	0.0045087	0.0028652
AC bonus squared	0.0000913	0.0000646	0.0001438	0.0000768
RC bonus	0.0050032	0.006936	0.045615	0.0079363
RC bonus squared	0.0002769	0.0003867	0.0031395	0.0004883

NOTE: SE is the standard error of the coefficient. Statistical significance: *** = 0.001, ** = 0.01, * = 0.1.

education level at the time of the reenlistment decision point to our data on reenlistment decisions, bonuses, and deployment histories.

Analysis Data File. We link the above data sources via Social Security numbers and dates of reenlistment to create a decision-level data file with over 2 million observations. To better understand the impact of our sample restrictions, Tables A.3 and A.4 compare the characteristics of our analytic sample to those of the universe of all service members at a reenlistment point.

The analytic file has higher RC enlistment rates and lower AC reenlistment rates than the full sample. This reflects the fact that our analysis is restricted to specialties in which a sufficient number of service members both reenlisted and joined the reserves.

Table A.3
Means of Key Variables for the Full and Analytic Samples, by Service: First Term

	Army		Navy		Marine Corps		Air Force	
	Analytic	Full	Analytic	Full	Analytic	Full	Analytic	Full
Reenlist	.300	.318	.384	.438	.309	.271	.465	.507
Join RC[a]	.312	.252	.160	.108	.090	.052	.175	.102
Leave force[b]	.388	.430	.456	.453	.601	.677	.360	.391
AC bonus pct.	.701		.721		.588		.579	
RC bonus pct.	.679		.292		.092		.230	
AC bonus amt.	7.657		10.632		10.603		12.312	
RC bonus amt.	6.594		4.416		1.907		3.026	
MOS 0	.324	.269	.123	.097	.393	.287	.184	.114
MOS 1	.067	.085	.093	.139	.010	.067	.065	.102
MOS 2	.099	.126	.097	.111	.072	.076	.053	.102
MOS 3	.076	.079	.081	.066	.000	.000	.078	.086
MOS 4	.018	.030	.008	.018	.009	.027	.046	.049
MOS 5	.136	.127	.125	.100	.152	.144	.221	.187
MOS 6	.111	.133	.321	.334	.087	.169	.266	.250
MOS 7	.016	.022	.057	.062	.012	.029	.028	.049
MOS 8	.149	.124	.095	.071	.178	.145	.058	.051
MOS 9	.002	.006	.000	.001	.087	.056	.001	.008

Table A.3—Continued

	Army		Navy		Marine Corps		Air Force	
	Analytic	Full	Analytic	Full	Analytic	Full	Analytic	Full
Deployed 1–8 months	.262	.268	.579	.562	.342	.356	.429	.408
Deployed 9+ months	.411	.386	.189	.179	.364	.341	.119	.113
Hostile deployed 1–8 months	.237	.240	.647	.608	.351	.358	.397	.369
Hostile deployed 9+ months	.357	.328	.058	.055	.236	.220	.087	.079
Male	.838	.840	.797	.829	.944	.940	.737	.752
Black	.207	.205	.265	.228	.119	.113	.184	.174
Hispanic	.126	.121	.140	.125	.161	.142	.064	.061
AFQT 3a	.305	.297	.283	.275	.278	.275	.315	.293
AFQT 2	.350	.377	.244	.322	.298	.368	.400	.442
AFQT 1	.046	.057	.017	.032	.029	.039	.039	.058
High school	.854	.844	.897	.887	.949	.948	.766	.792
Some college	.045	.057	.017	.021	.013	.014	.232	.205
Single	.632	.613	.633	.619	.565	.552	.524	.513
Single no dependents	.485	.468	.502	.505	.531	.518	.469	.459
Fast promotion	.072	.097	.168	.216	.220	.264	.012	.025
Years of service at decision	3.865	4.033	4.213	4.229	4.017	4.188	4.447	4.529
1999	.107	.099	.096	.091	.108	.093	.122	.096

Table A.3—Continued

	Army		Navy		Marine Corps		Air Force	
	Analytic	Full	Analytic	Full	Analytic	Full	Analytic	Full
2000	.102	.093	.096	.094	.114	.092	.143	.102
2001	.085	.096	.098	.095	.106	.100	.065	.077
2002	.084	.086	.089	.100	.106	.099	.077	.091
2003	.082	.094	.099	.095	.082	.098	.061	.076
2004	.123	.124	.122	.114	.093	.099	.099	.099
2005	.125	.121	.116	.115	.105	.103	.116	.115
2006	.111	.106	.107	.106	.100	.108	.099	.111
2007	.102	.095	.108	.100	.122	.109	.118	.122

NOTE: MOS = military occupation specialty.

[a] "Join RC" is the fraction of AC service members at a reenlistment point who transition to the RC within two years of leaving the AC.

[b] "Leave Force" is the fraction of AC service members at a reenlistment point who leave the AC and do not join the RC in the following two years.

Table A.4
Means of Key Variables for the Full and Analytic Samples, by Service: Second Term Plus

	Army		Navy		Marine Corps		Air Force	
	Analytic	Full	Analytic	Full	Analytic	Full	Analytic	Full
Reenlist	.549	.555	.457	.537	.522	.621	.624	.631
Join RC[a]	.162	.082	.182	.042	.131	.015	.160	.036
Leave force[b]	.289	.364	.362	.422	.347	.364	.217	.333
AC bonus pct.	.701		.677		.397		.678	
RC bonus pct.	.591		.249		.052		.205	
AC bonus amt.	8.485		14.862		6.581		12.652	
RC bonus amt.	5.510		3.723		.685		2.371	
MOS 0	.259	.244	.073	.093	.206	.175	.135	.087
MOS 1	.043	.055	.107	.162	.012	.073	.054	.092
MOS 2	.055	.098	.082	.102	.057	.086	.044	.084
MOS 3	.084	.085	.175	.085	.000	.001	.076	.083
MOS 4	.019	.038	.010	.034	.009	.036	.029	.043
MOS 5	.223	.193	.210	.158	.424	.297	.345	.287
MOS 6	.122	.135	.197	.253	.085	.156	.238	.222
MOS 7	.009	.019	.045	.055	.011	.024	.024	.043
MOS 8	.186	.130	.102	.058	.122	.110	.054	.054
MOS 9	.000	.003	.000	.000	.075	.042	.000	.004
Deployed 1–8 months	.306	.305	.428	.436	.422	.426	.436	.420

Table A.4—Continued

	Army		Navy		Marine Corps		Air Force	
	Analytic	Full	Analytic	Full	Analytic	Full	Analytic	Full
Deployed 9+ months	.371	.339	.128	.162	.191	.231	.148	.137
Hostile deployed 1–8 months	.265	.245	.389	.395	.235	.282	.383	.323
Hostile deployed 9+ months	.268	.223	.033	.035	.078	.100	.072	.059
Male	.810	.845	.787	.873	.929	.940	.738	.829
Black	.333	.326	.288	.212	.220	.223	.212	.191
Hispanic	.112	.091	.132	.094	.180	.121	.057	.045
AFQT 3a	.303	.280	.258	.226	.303	.275	.330	.279
AFQT 2	.262	.312	.327	.401	.291	.355	.394	.428
AFQT 1	.022	.041	.044	.069	.022	.038	.032	.058
High school	.861	.800	.903	.852	.930	.873	.636	.513
Some college	.038	.131	.034	.080	.039	.085	.364	.487
Single	.364	.283	.415	.284	.280	.201	.306	.220
Single no dependents	.209	.150	.290	.182	.202	.120	.210	.126
Fast promotion	.041	.085	.031	.117	.200	.169	.002	.085
Years of service at decision	6.725	10.633	7.712	13.050	7.851	12.734	8.809	15.108
1999	.114	.118	.129	.100	.112	.086	.150	.111
2000	.123	.114	.132	.101	.145	.095	.163	.115

Table A.4—Continued

	Army		Navy		Marine Corps		Air Force	
	Analytic	Full	Analytic	Full	Analytic	Full	Analytic	Full
2001	.118	.110	.142	.102	.137	.094	.067	.090
2002	.087	.100	.087	.104	.113	.094	.069	.098
2003	.070	.080	.073	.097	.077	.091	.068	.102
2004	.088	.097	.080	.099	.072	.096	.079	.096
2005	.120	.108	.087	.097	.069	.103	.102	.096
2006	.108	.092	.106	.107	.101	.105	.086	.093
2007	.095	.091	.106	.100	.111	.131	.119	.103

[a] "Join RC" is the percentage of AC service members at a reenlistment point who transition to the RC within two years of leaving the AC.

[b] "Leave Force" is the percentage of AC service members at a reenlistment point who leave the AC and do not join the RC in the following two years.

Since the RC rates are quite low in the overall sample, the RC enlistment rate in specialties with a positive number of service members choosing to join the reserves should be higher than in the general population. The analytic file also has lower average years of service. This is associated with the analytic sample having higher percentages of service members in the first term, in combat arms, single, without dependents, and with lower levels of education. We estimate separate models for first-term and second-term-plus service members, thereby taking a number of these sample differences into account.

Issues in Estimating Bonus Effects

As mentioned in the text, there are several limitations to our application of the conditional logit model. These include functional form restrictions, including but not limited to the independence of irrelevant alternatives (IIA) assumption, and complications arising from bonus caps, deployment-related bonuses, stop-loss, reverse causality, future bonuses, and the unavailability of vacancies in reserve units.

First, the random utility model with GEV errors places a number of restrictions on the substitution patterns across alternatives in the model. For example, the model embeds the IIA assumption, which implies that the ratio of probabilities of any two alternatives is independent of the other choices in the choice set. That is, under IIA adding or dropping a choice from the choice set will not affect the ratio of probabilities of any two choices that remain in the set. We performed tests of the IIA assumption on a number of specifications, and we rejected the null in all cases, thus suggesting the presence of an IIA violation.

To address concerns over restrictions on substation patterns embedded in the conditional logit model, we replicated a number of key specifications with models that relax those patterns. First, we ran a number of specifications using a conditional probit model, which does not require the IIA assumption, and the results are nearly identical to those reported in the text. Second, we replicated a number of models using a nested logit model, where the nests were {AC} and {RC, C}. The nested logit model allows for a variety of substitution patterns across alternatives and relaxes the IIA assumption for choices across nests. The results with the nested logit model were mixed. The overall pattern of positive own-effects and negative cross-effects of bonuses remained unchanged. However, in general, the coefficient estimates for the reserve bonus effect were somewhat larger in magnitude than, and those for the active bonus were similar to, those we obtained with the conditional logit model. Moreover, while most of our nested logit models produced reasonable estimates, several models failed to converge, while others implied correlations that are inconsistent with any random utility model. This could be due to the strong functional form assumptions embedded in the nested logit model.

Second, bonus caps might bias the bonus effect down at higher levels of the bonus but not at lower levels. This can be explained as follows. Our RC bonus variable is an average of the bonuses received by service members in a given specialty who join the RC. They may stay in their own specialty or enter another, and of course bonus offerings can change over time. If a bonus is at its ceiling, the bonus cannot increase further, and if an individual is on his supply curve, his willingness to enlist in the RC in the specialty offering will be no higher than that corresponding to the capped bonus. As more specialties offer a bonus at the cap, our bonus variable increases, but the reenlistment probability will increase at a lower rate than before.[1]

The AC selective reenlistment bonus cap was $45,000 in 1999 and increased to $90,000 in 2006. In our sample, the average AC reenlistment bonus was less than $12,500 in all services. The cap for the reserve affiliation bonus was $6,000 prior to 2006, and in 2006 it increased to $10,000 for a three-year enlistment and $20,000 for a six-year enlistment. The caps for a reserve prior service enlistment bonus before 2006 were $5,000 for a three-year enlistment and $10,000 for a six-year enlistment, and in 2006 they increased to $7,500 and $15,000, respectively.[2] The average RC bonus in the sample was $6,162 in the Army, $4,228 in the Navy, $1,811 in the Marine Corps, and $2,798 in the Air Force, though as shown in Chapter Two (Figure 2.4) all of the RC paid low bonuses averaging less than $2,000 until 2005, when the average amounts rose rapidly and stayed at high levels through at least 2008. The Navy, Marine Corps, and Air Force average bonuses are well below both the affiliation bonus cap and the enlistment bonus cap. The Army bonus was closer to the caps.

To account for possible bonus cap effects, we ran models that allow the relationship between bonus amounts and reenlistment decisions to be quadratic.[3] We find that bonus effects decrease as bonus amounts increase. The decrease is largely negligible for the AC but noticeable for the RC. Moreover, as expected, our estimated RC bonus effect peaks at a lower level for the Army than it does for other services.

---

[1]   Suppose a service member is considering two types of RC slots, and one type has bonus b_1 and the other type has bonus b_2. The fraction of type one slots is θ. First assume both bonuses are below the bonus caps. In a simple linear supply model, the expected probability of RC enlistment is $\theta e_1 + (1-\theta)e_2 = \theta \beta b_1 + (1-\theta)\beta b_2 = \beta \hat{b}$, where $\hat{b}$ is our bonus variable. In this case, the estimate of the bonus effect will not be biased. Now assume that b_1 has been assigned to slots that are hard to fill and is at the cap, with the result that $\bar{b}_1$ produces enlistment $\bar{e}_1$. Because the slots are hard to fill, we can assume $\bar{e}_1 < e_2$ and of course $\bar{b}_1 > b_2$. In this case, the expected RC enlistment is $\theta \bar{e}_1 + (1-\theta)e_2 = \theta \beta \bar{b}_1 + (1-\theta)\beta b_2$ and all of the bonus variation comes from the bonus that is not at the cap. Finally, consider the situation as the bonus variable increases and nears the cap. This occurs because the fraction of hard-to-fill slots is increasing. But as θ increases, the expected enlistment *decreases* while the expected bonus increases.

[2]   Hattiangadi et al. (2006) report that AC bonuses offerings as of 2005 were consistent with these limits.

[3]   We also ran models with a linear bonus specification. The quadratic specification outperformed the linear specification in that the quadratic terms were statistically significant and the log likelihood was significantly higher. The coefficients on all other variables were nearly identical between the linear and quadratic bonus specifications.

Third, our bonus variable for the AC includes selective reenlistment bonuses and deployment bonuses, the latter being offered by the Army.[4] Selective reenlistment bonuses are offered at the discretion of the service by three-digit occupational specialty and zone (length of service 21 months to 6 years, from 6 to 10 years, and from 10 to 14 years). The bonus amount is the product of the service member's basic pay at the time of reenlistment, years of service obligated, and a factor called the bonus multiplier, which ranges from 1 to 6 and can be set in half steps.

The Army has implemented several variants of reenlistment bonuses. In 2008 the Army changed from this formula and set bonus amounts based on an award table defined in terms of pay grade and years of service obligated, and bonuses were paid as a lump sum at the time of reenlistment rather than half up front and the remainder on anniversary payments during the term of service. Further, beginning in June 2003 the Army offered bonuses for reenlistment during a deployment to Iraq, Afghanistan, Kuwait, or assignment to Korea. These bonuses were a $5,000 lump sum, did not depend on occupational specialty, and focused on soldiers in pay grades E-4, E-5, and E-6. Soldiers who were also eligible for a selective reenlistment bonus could opt for the larger of the two bonuses. In the following fiscal years, the Army had a Selective Reenlistment Bonus–Deployed Program. This program allowed soldiers to reenlist outside of their reenlistment window, which normally begins 18 months prior to the expiration of term-of-service date. The program offered a bonus to soldiers, regardless of occupational specialty, with 17 months to 6 years of service, and to soldiers with 6 to 10 years of service, with bonus multipliers of 1.5 and 1.0 times basic pay, respectively. By reenlisting in a combat zone, a soldier would receive the full amount of the bonus free of federal income tax. Soldiers who also qualified for a selective reenlistment bonus in their specialty at the time they were deployed could choose to take the larger of the two. Finally, in 2008 the Army offered for the first time a "reset" bonus for soldiers in designated specialties with more than 24 months remaining on their current service contract and who had returned from a deployment within the past 120 days. The purpose was to "lock in" experienced personnel in selected specialties so they would be available for the next deployment.

From 2004 to 2009, the Army implemented a stop-loss policy. The policy applied throughout the Army to AC and RC soldiers. Under stop-loss, a soldier with an ETS date within 90 days of the date of unit deployment was required to deploy and remain

---

[4] There are other types of pays related to deployment. Beginning in 2004, the Army offered a bonus of $1,000 per month to soldiers whose tours in Iraq or Afghanistan were *extended* beyond 12 months. Of the $1,000, $200 came as Extra Hardship Duty Pay and $800 came as Assignment Incentive Pay. The Marine Corps also used Assignment Incentive Pay. For example, in fiscal years 2007–2009 the Marine Corps offered bonuses to AC marines to extend their current term of service to complete a deployment to the Middle East, with $3,000 for a seven-month deployment and $6,000 for a 12-month deployment. In addition, the Army also offered a bonus for extending the current term of service and retraining in a shortage occupation under its BEAR program (bonus extension and retraining). We do not include any of these bonuses, which are for extensions, because we limit our analysis to reenlistment decisions.

in the military until 90 days after returning from deployment. A soldier could reenlist during this period but could not leave. For soldiers who wanted to leave, stop-loss imposed an involuntary extension of service. The incentive to reenlist during deployment was greater if a reenlistment bonus was offered, especially because reenlistment in a Combat Zone Tax Exclusion Zone meant that the entire bonus was nontaxable. These particular conditions (reenlistment in a Combat Zone Tax Exclusion Zone area under stop-loss) tend to impart an upward bias to the effect of a bonus on reenlistment (Hosek and Martorell, 2009).

Fourth, in order to adopt a causal interpretation of our point estimates, we must assume that, conditional upon the covariates included in our model, deployment history and bonus offers are both uncorrelated with the error terms in the model. Lyle (2006), Savych (2008), and Engel, Gallagher, and Lyle (2010) each created an instrument for individual deployment, namely, one-third or more of the individual's unit members were deployed in the prior year. The coefficient estimates these studies obtained when using the instrument were quite similar to those obtained when using observed individual deployment. Also, each study ran Hausman tests and in all cases could not reject the null hypothesis that the error term was orthogonal to deployment. Given these results, we treat deployment as exogenous given controls for MOS and year fixed effects. We control for one-digit occupational specialty due to the computational burden of controlling for it at the three-digit level.

Yet even with occupation and year fixed effects, bonuses may be endogenous with respect to reenlistment decisions. First, there may be reverse causality. Bonus setters may increase bonuses when reenlistment conditions are poor, and this imparts a negative bias to the estimated effect of bonuses on recruiting or reenlistment. Second, there may be unobserved factors that affect the bonus estimates. For example, if reenlistment counselors work harder when the military needs to retain more service members, and at the same time the military increases reenlistment bonuses, the unobserved counselor effort would bias up the estimated bonus effect. While controlling for occupation fixed effects mitigates selection bias, there still may be some reverse causality and omitted variable bias. The sign of the bias may be negative or positive, and the extent of the bias is unknown.

Fifth, reenlistment timing may depend on future bonuses. If service members know a future bonus will be greater than the current bonus, they may delay reenlistment to the future. If the future bonus is omitted from the regression specification, this timing behavior would tend to bias down the effect of the current bonus on current reenlistment. However, we expect the bias stemming from strategic timing of reenlistment decisions is likely to be small. The extent to which service members can accurately forecast future bonuses is unclear. In general, the bonus setter will increase the bonus if there is a shortfall of reenlistments relative to reenlistment target increases. To forecast the shortfall requires considering both the future target and expected future reenlistment. Further, the bonus setter can make the same forecast and may incorporate it when setting the current bonus; rather than allowing a future shortfall to materialize, the bonus setter will increase the current bonus. The service member might factor the bonus setter's dynamic responsiveness into the forecast of the future bonus. Because of the complexity of the forecast and possible response of the bonus setter, we expect the bias from reenlistment timing behavior to be small.

Finally, our model assumes that service members nearing a reenlistment decision point have the option to join a reserve unit. In reality, some reserve units may not have vacancies. Reserve bonuses are set nationally, and if some local units do not have openings in specialties offering a bonus, it would not be feasible for an individual to join the unit. This would tend to bias the bonus effect down. However, we do not expect this to be a large problem, for two reasons. Many service members change specialties when they leave an AC and join an RC, and this flexibility increases the chance an opening will be available. Further, reserve units have substantial personnel turnover, which implies that a specialty that does not have a current opening may have one within six months (most AC leavers who join the reserve do so within six months) and is even more likely to have one within two years (the length of the window we use to observe reserve enlistment). Lippiatt and Polich (2010) studied reserve units that deployed and found "between 25 and 40 percent of personnel who were assigned to the unit 12 months before mobilization had left the unit during the subsequent year." Also, "of all the soldiers who actually deployed with those units, 40 to 50 percent were 'new arrivals' who had been in the unit less than one year." Although Lippiatt and Polich did not study units not preparing to deploy, such units also may have substantial turnover.

First-Term Regression Results

Table C.1, which begins on the following page, presents our first-term regression results.

Table C.1
First-Term Regression Results

	Army				Navy				Marine Corps				Air Force			
	Choose AC		Choose RC		Choose AC		Choose RC		Choose AC		Choose RC		Choose AC		Choose RC	
	Est	SE	Est	SE	Est	SE	Est	SE	Est	SE	Est	SE	Est	SE	Est	SE
AC bonus	.034	.0030	.034	.0030	.064	.0024	.064	.0024	.090	.0037	.090	.0037	.016	.0024	.016	.0024
AC bonus squared	.000	.0001	.000	.0001	-.001	.0001	-.001	.0001	-.002	.0001	-.002	.0001	.000	.0001	.000	.0001
RC bonus	.037	.0033	.037	.0033	.046	.0082	.046	.0082	.045	.0332	.045	.0332	.044	.0084	.044	.0084
RC bonus squared	-.002	.0002	-.002	.0002	-.002	.0004	-.002	.0004	-.001	.0023	-.001	.0023	-.002	.0005	-.002	.0005
Male	.357	.0187	.068	.0187	.273	.0210	-.095	.0260	.128	.0450	-.161	.0607	.151	.0211	-.079	.0272
Black	.580	.0187	.217	.0173	.843	.0191	.362	.0253	.775	.0326	.259	.0475	.579	.0227	.065	.0312
Hispanic	.189	.0172	.312	.0188	.354	.0237	.404	.0287	.269	.0292	.324	.0380	.197	.0339	.135	.0434
AFQT 3a	.358	.0206	-.187	.0161	-.252	.0195	-.170	.0250	-.046	.0263	-.058	.0363	-.114	.0221	.067	.0297
AFQT 2	.220	.0171	-.241	.0162	-.384	.0220	-.198	.0276	-.090	.0276	-.055	.0367	-.321	.0225	-.013	.0296
AFQT 1	.052	.0176	-.222	.0312	-.654	.0649	-.489	.0826	-.219	.0722	-.109	.0880	-.614	.0467	-.181	.0569
High school	.139	.0352	-.134	.0200	-.129	.0279	-.043	.0366	-.120	.0542	-.186	.0721	.183	.1610	.048	.2147
Some college	-.585	.0221	-.305	.0346	-.324	.0679	.006	.0849	-.469	.1149	-.114	.1350	-.124	.1627	.110	.2165
Single	-.220	.0389	.112	.0195	-.049	.0255	.069	.0340	-.354	.0584	.200	.0740	-.079	.0373	.202	.0487
Single no dependents	-.391	.0194	.102	.0182	-.404	.0248	.016	.0321	-.407	.0582	-.171	.0723	-.391	.0375	-.018	.0482
Fast promotion	1.711	.0192	.334	.0315	.821	.0230	.379	.0307	.807	.0271	.572	.0338	.735	.0930	1.384	.0972

Table C.1—Continued

	Army				Navy				Marine Corps				Air Force			
	Choose AC		Choose RC		Choose AC		Choose RC		Choose AC		Choose RC		Choose AC		Choose RC	
	Est	SE	Est	SE	Est	SE	Est	SE	Est	SE	Est	SE	Est	SE	Est	SE
Years of service at decision	-.322	.0270	-.002	.0078	.003	.0137	.292	.0169	-3.413	.0381	.076	.0337	-.251	.0098	.073	.0121
MOS 1	.626	.0090	-.132	.0267	.078	.0378	-.003	.0448	3.071	.1162	.901	.1304	.539	.0390	.916	.0466
MOS 2	.343	.0270	-.020	.0219	.272	.0353	.122	.0420	.673	.0420	.130	.0628	.507	.0449	.755	.0554
MOS 3	.976	.0241	-.085	.0272	.376	.0402	-.289	.0500					.865	.0382	.865	.0486
MOS 4	.595	.0265	.273	.0465	1.119	.0877	.583	.1038	1.571	.1070	1.023	.1497	-.063	.0420	.331	.0528
MOS 5	.937	.0496	-.024	.0228	.878	.0318	.162	.0397	1.066	.0367	.483	.0479	.891	.0298	.554	.0384
MOS 6	.337	.0234	.152	.0203	.211	.0271	.025	.0317	.981	.0399	.530	.0518	.651	.0254	.803	.0328
MOS 7	.700	.0236	.068	.0489	.355	.0403	-.011	.0497	.874	.0938	.851	.1111	.681	.0533	1.103	.0629
MOS 8	.690	.0514	-.180	.0201	.536	.0362	-.180	.0488	.645	.0336	.230	.0442	.546	.0392	.580	.0500
MOS 9	-.284	.0211	.330	.1215					-.693	.0502	-.077	.0528	1.076	.3540	.177	.5829
1999	.379	.1777	.598	.0318	.261	.0392	1.007	.0532	.559	.0749	.846	.1110	-.245	.0417	.128	.0533
2000	.427	.0346	.693	.0325	.394	.0390	.998	.0537	.548	.0736	.863	.1110	-.123	.0370	.252	.0479
2001	.354	.0340	.551	.0336	.579	.0389	.845	.0553	.691	.0745	.853	.1116	-.160	.0434	.388	.0555
2002	.428	.0348	.526	.0335	.791	.0394	.814	.0570	.842	.0741	.718	.1125	-.549	.0414	-.102	.0534
2003	-.013	.0348	.137	.0324	.407	.0387	.724	.0549	.531	.0760	.273	.1167	-.343	.0447	.065	.0586
2004	.087	.0357	-.023	.0286	.267	.0375	.612	.0531	.504	.0726	.411	.1141	-.450	.0385	.088	.0499

Table C.1—Continued

	Army				Navy				Marine Corps				Air Force			
	Choose AC		Choose RC		Choose AC		Choose RC		Choose AC		Choose RC		Choose AC		Choose RC	
	Est	SE	Est	SE	Est	SE	Est	SE	Est	SE	Est	SE	Est	SE	Est	SE
2005	-.100	.0315	-.034	.0276	.199	.0373	.449	.0544	.386	.0697	.508	.1120	-.333	.0348	-.066	.0469
2006	-.312	.0299	.057	.0278	.218	.0383	.290	.0565	.450	.0692	.581	.1125	-.233	.0363	-.019	.0479
2007	.042	.0312	.215	.0286	.116	.0393	.295	.0574	-.002	.0651	.248	.1158	-.220	.0346	-.058	.0460
Deployed 1–8 months	.207	.0313	-.040	.0292	.381	.0368	.113	.0509	.009	.0382	.099	.0535	.433	.0375	.237	.0517
Deployed 9+ months	.190	.0270	.152	.0344	.496	.0434	.187	.0599	-.117	.0458	.186	.0632	.655	.0546	.295	.0741
Hostile deployed 1–8 months	.193	.0325	.067	.0293	-.257	.0340	-.065	.0479	.017	.0364	-.053	.0511	-.334	.0374	.013	.0514
Hostile deployed 9+ months	-.502	.0269	.025	.0358	-.366	.0521	-.065	.0705	-.095	.0509	-.218	.0724	-.612	.0603	.108	.0798
Constant	-.423	.0343	-.605	.0502	-1.491	.0778	-3.057	.1000	11.578	.1797	-2.906	.1973	.946	.1701	-2.065	.2271
N	489,642				263,811				183,246				241,374			
Log-likelihood	-166,278.46				-84,241.062				-44,720.22				-79,304.52			

Second-Term Regression Results

Table D.1, which begins on the following page, presents our second-term regression results.

Table D.1
Second-Term Regression Results

	Army				Navy				Marine Corps				Air Force			
	Choose AC		Choose RC		Choose AC		Choose RC		Choose AC		Choose RC		Choose AC		Choose RC	
	Est	SE	Est	SE	Est	SE	Est	SE	Est	SE	Est	SE	Est	SE	Est	SE
AC bonus	.054	.0028	.054	.0028	.046	.0035	.046	.0035	.053	.0098	.053	.0098	.032	.0034	.032	.0034
AC bonus squared	-.001	.0001	-.001	.0001	-.001	.0001	-.001	.0001	-.001	.0004	-.001	.0004	-.001	.0001	-.001	.0001
RC bonus	.018	.0050	.018	.0050	.066	.0127	.066	.0127	-.041	.0955	-.041	.0955	.019	.0132	.019	.0132
RC bonus squared	-.001	.0003	-.001	.0003	-.003	.0007	-.003	.0007	.000	.0059	.000	.0059	-.001	.0008	-.001	.0008
Male	.580	.0201	.165	.0265	.331	.0340	.012	.0415	-.035	.1253	-.386	.1625	.410	.0324	.075	.0418
Black	.385	.0178	.080	.0240	.427	.0330	.218	.0418	.713	.0842	.173	.1275	.307	.0336	-.137	.0454
Hispanic	.214	.0246	.362	.0318	.269	.0417	.367	.0504	.273	.0869	.573	.1166	.219	.0562	.080	.0727
AFQT 3a	-.059	.0177	.012	.0239	-.197	.0346	-.150	.0437	-.070	.0776	-.071	.1141	-.108	.0342	-.036	.0451
AFQT 2	-.122	.0193	.061	.0255	-.340	.0359	-.123	.0442	-.120	.0828	.163	.1163	-.252	.0349	-.096	.0454
AFQT 1	-.127	.0524	.259	.0641	-.712	.0865	-.094	.0935	.105	.2329	.471	.2995	-.587	.0770	-.425	.0991
High school	.194	.0237	.239	.0343	.181	.0548	-.019	.0672	.174	.1750	.535	.2951	1.713	.9357	.697	1.1359
Some college	.154	.0451	.407	.0580	.211	.0942	.338	.1091	.125	.2429	1.145	.3540	1.391	.9358	.912	1.1361
Single	-.139	.0204	.146	.0271	-.099	.0424	.105	.0520	-.079	.1206	.421	.1584	.011	.0450	.240	.0566
Single no dependents	.125	.0243	.204	.0314	-.166	.0463	-.067	.0558	-.213	.1353	-.141	.1755	-.123	.0502	-.107	.0630
Fast promotion	1.407	.0483	.858	.0652	.291	.0833	.265	.0997	1.174	.0946	.882	.1301	1.459	.2927	1.702	.3265

Table D.1—Continued

	Army				Navy				Marine Corps				Air Force			
	Choose AC		Choose RC		Choose AC		Choose RC		Choose AC		Choose RC		Choose AC		Choose RC	
	Est	SE	Est	SE	Est	SE	Est	SE	Est	SE	Est	SE	Est	SE	Est	SE
Years of service at decision	.079	.0035	.142	.0044	.282	.0075	.201	.0089	.242	.0204	.285	.0274	.224	.0072	.189	.0090
MOS 1	.275	.0381	.169	.0491	-.736	.0685	-.097	.0804	.489	.3800	.883	.4429	.567	.0645	.741	.0801
MOS 2	.236	.0343	.238	.0441	-.518	.0676	-.129	.0826	-.132	.1635	.406	.2075	.343	.0843	.833	.0978
MOS 3	.457	.0302	-.007	.0406	-.247	.0588	-.312	.0753					.838	.0582	.730	.0742
MOS 4	.347	.0570	.369	.0720	.792	.1674	1.066	.1845	.158	.3816	.971	.4294	.107	.0771	.483	.0942
MOS 5	.536	.0241	.071	.0318	.375	.0559	.042	.0718	.037	.0924	-.239	.1318	.775	.0420	.333	.0555
MOS 6	.314	.0253	.182	.0330	-.381	.0575	.065	.0710	.136	.1304	.381	.1747	.738	.0413	.585	.0551
MOS 7	.610	.0849	.742	.1004	.085	.0802	.440	.0949	-.145	.3171	.285	.4107	.649	.0901	.979	.1066
MOS 8	.242	.0225	-.050	.0305					-.173	.1142	-.098	.1639	.795	.0652	.633	.0834
MOS 9	2.049	.6123	1.666	.6829	-.156	.0656	-.128	.0841	-.581	.1336	-.839	.2133				
1999	.335	.0381	.812	.0519	.368	.0686	.621	.0899	.952	.1908	.519	.2486	.475	.0602	-.012	.0761
2000	.481	.0376	.830	.0522	.415	.0678	.628	.0893	.808	.1841	.661	.2392	.393	.0550	.089	.0716
2001	.423	.0375	.947	.0520	.450	.0672	.686	.0889	1.094	.1867	.703	.2423	.391	.0671	.345	.0862
2002	.523	.0393	.858	.0549	.709	.0728	.643	.0967	1.233	.1929	.770	.2498	.312	.0657	.189	.0865
2003	.139	.0406	.548	.0555	.413	.0746	.500	.0980	1.153	.2005	.669	.2622	.358	.0680	.273	.0888
2004	.192	.0373	.306	.0537	.299	.0734	.249	.0987	1.032	.2015	.603	.2716	.081	.0622	.250	.0801

Table D.1—Continued

	Army				Navy				Marine Corps				Air Force			
	Choose AC		Choose RC		Choose AC		Choose RC		Choose AC		Choose RC		Choose AC		Choose RC	
	Est	SE	Est	SE	Est	SE	Est	SE	Est	SE	Est	SE	Est	SE	Est	SE
2005	.184	.0335	.277	.0509	.218	.0722	.114	.0993	.359	.1965	-.012	.2720	-.293	.0558	-.013	.0729
2006	.038	.0334	.233	.0518	.127	.0700	.001	.0965	.142	.1877	-.181	.2551	-.135	.0589	-.119	.0787
2007	-.002	.0342	.182	.0533	.030	.0718	.143	.0965	.737	.1760	.390	.2576	-.362	.0543	-.158	.0721
Deployed 1–8 months	.353	.0265	.123	.0345	.390	.0448	.031	.0569	.554	.0814	.031	.1186	.227	.0449	.232	.0576
Deployed 9+ months	.455	.0289	.236	.0377	.761	.0633	.109	.0823	.462	.1215	.037	.1799	.365	.0584	.252	.0754
Hostile deployed 1–8 months	.151	.0264	-.034	.0350	-.036	.0455	-.033	.0585	-.017	.0937	.165	.1372	.081	.0444	.029	.0568
Hostile deployed 9+ months	-.294	.0318	-.022	.0429	-.075	.0985	-.043	.1332	-.101	.1680	.106	.2504	.060	.0735	.183	.0939
Constant	-1.738	.0533	-2.852	.0701	-2.975	.1172	-2.668	.1456	-3.154	.3250	-4.232	.4688	-3.683	.9401	-3.365	1.1415
N	323,217				93,690				16,755				126,366			
Log-likelihood	-101,723.72				-30,077.39				-4,985.41				-36,920.5			

Bibliography

Arkes, Jeremy, and M. Rebecca Kilburn (2005). *Modeling Reserve Recruiting: Estimate of Enlistments.* Santa Monica, Calif.: RAND Corporation, MG-202-OSD. As of February 18, 2011: http://www.rand.org/pubs/monographs/MG202.html

Asch, Beth, and David Loughran (2005). *Reserve Recruiting and the College Market: Is a New Educational Benefit Needed?* Santa Monica, Calif.: RAND Corporation, TR-127-OSD. As of February 18, 2011: http://www.rand.org/pubs/technical_reports/TR127.html

Asch, Beth, Paul Heaton, James Hosek, Francisco Martorell, Curtis Simon, and John Warner (2010). *Cash Incentives and Military Enlistment, Attrition, and Reenlistment.* Santa Monica, Calif.: RAND Corporation, MG-950-OSD. As of February 18, 2011: http://www.rand.org/pubs/monographs/MG950.html

Boskin, Michael (1974). "A Conditional Logit Model of Occupational Choice." *Journal of Political Economy*, Vol. 82, No. 2, pp. 389–398.

Engel, Rozlyn C., Luke B. Gallagher, and David S. Lyle (2010). "Military Deployments and Children's Academic Achievement: Evidence from Department of Defense Education Activity Schools." *Economics of Education Review*, Vol. 29, No. 1, pp. 73–82.

Friedman, Joseph (1981). "A Conditional Logit Model of the Role of Local Public Services in Residential Choice." *Urban Studies,* Vol. 18, No. 3, pp. 347–357.

Hattiangadi, Anita U., and Ann D. Parcell, with David Gregory and Ian D. MacLeod (2006). *SelRes Attrition and the Selected Reserve Incentive Program in the Marine Corps Reserve.* Alexandria, Va.: Center for Naval Analyses, CRM D0013618.A2/Final. As of February 18, 2011: http://www.cna.org/research/2006/selres-attrition-selected-reserve-incentive-marine

Hausman, Jerry, and David Wise (1977). "Social Experimentation, Truncated Distributions, and Efficient Estimation." *Econometrica*, Vol. 45, No. 4, pp. 919–938.

Hosek, James, and Francisco Martorell (2009). *How Have Deployments During the War on Terrorism Affected Reenlistment?* Santa Monica, Calif.: RAND Corporation, MG-873-OSD. As of February 18, 2011: http://www.rand.org/pubs/monographs/MG873.html

Hosek, James, and Christine Peterson (1985). *Reenlistment Bonuses and Retention Behavior.* Santa Monica, Calif.: RAND Corporation, R-3199-MIL. As of February 18, 2011: http://www.rand.org/pubs/reports/R3199.html

Hosek, James, and Mark Totten (2002). *Serving Away From Home: How Deployments Influence Reenlistment.* Santa Monica, Calif.: RAND Corporation, MR-1594-OSD. As of February 18, 2011: http://www.rand.org/pubs/monograph_reports/MR1594.html

Kostiak, Peter F., and James E. Grogan (1987). *Enlistment Supply into the Naval Reserve*. Alexandria, Va.: Center for Naval Analyses, CRM 87-239. As of February 18, 2011:
http://www.cna.org/research/1987/enlistment-supply-naval-reserve

Lippiatt, Thomas F., and J. Michael Polich (2010). *Reserve Component Unit Stability: Effects on Deployability and Training*. Santa Monica, Calif.: RAND Corporation, MG-954-OSD. As of February 18, 2011:
http://www.rand.org/pubs/monographs/MG954.html

Long, Bridget (2004). "How Have College Choices Changed Over Time? An Application of the Conditional Logistic Choice Model." *Journal of Econometrics*, Vol. 121, No. 1, pp. 271–296.

Lyle, David S. (2006). "Using Military Deployments and Job Assignments to Estimate the Effect of Parental Absences and Household Relocations on Children's Academic Achievement." *Journal of Labor Economics*, Vol. 24, No. 2, pp. 319–350.

McFadden, Daniel (1974). *Conditional Logit Analysis of Qualitative Choice Behavior*. In *Frontiers in Econometrics*, Paul Zarembeka, ed. New York: Academic Press.

Montmarquette, Claude, Kathy Cannings, and Sophie Mahseredjian (2002). "How Do Young People Choose College Majors?" *Economics of Education Review*, Vol. 21, No. 6, pp. 543–556.

Savych, Bogdan (2008). *Effects of Deployments on Spouses of Military Personnel*. Santa Monica, Calif.: RAND Corporation, RGSD-233. As of February 18, 2011:
http://www.rand.org/pubs/rgs_dissertations/RGSD233.html

Train, Kenneth (2009). *Discrete Choice Methods with Simulation*. 2nd edition. Cambridge, UK: Cambridge University Press.

Wooldridge, Jeffrey (2001). *Introductory Econometrics: A Modern Approach*. 2nd edition, South-Western College Publishing.